NISTIR 7916

Proceedings of the Cybersecurity in Cyber-Physical Systems Workshop, April 23 – 24, 2012

Editor:
Tanya Brewer
Computer Security Division
Information Technology Laboratory

http://dx.doi.org/10.6028/NIST.IR.7916

February 2013

U.S. Department of Commerce
Rebecca M. Blank, Acting Secretary of Commerce and Deputy Secretary of Commerce

National Institute of Standards and Technology
Patrick D. Gallagher, Under Secretary of Commerce for Standards and Technology and Director

Reports on Computer Systems Technology

The Information Technology Laboratory (ITL) at the National Institute of Standards and Technology (NIST) promotes the U.S. economy and public welfare by providing technical leadership for the Nation's measurement and standards infrastructure. ITL develops tests, test methods, reference data, proof of concept implementations, and technical analyses to advance the development and productive use of information technology. ITL's responsibilities include the development of management, administrative, technical, and physical standards and guidelines for the cost-effective security and privacy of other than national security-related information in Federal information systems.

Authority

This publication has been developed by NIST to further its statutory responsibilities under the Federal Information Security Management Act (FISMA), Public Law (P.L.) 107-347. NIST is responsible for developing information security standards and guidelines, including minimum requirements for Federal information systems, but such standards and guidelines shall not apply to national security systems without the express approval of appropriate Federal officials exercising policy authority over such systems. This guideline is consistent with the requirements of the Office of Management and Budget (OMB) Circular A-130, Section 8b(3), *Securing Agency Information Systems*, as analyzed in Circular A-130, Appendix IV: *Analysis of Key Sections*. Supplemental information is provided in Circular A-130, Appendix III, *Security of Federal Automated Information Resources*.

Nothing in this publication should be taken to contradict the standards and guidelines made mandatory and binding on Federal agencies by the Secretary of Commerce under statutory authority. Nor should these guidelines be interpreted as altering or superseding the existing authorities of the Secretary of Commerce, Director of the OMB, or any other Federal official. This publication may be used by nongovernmental organizations on a voluntary basis and is not subject to copyright in the United States. Attribution would, however, be appreciated by NIST.

Certain commercial entities, equipment, or materials may be identified in this document in order to describe an experimental procedure or concept adequately. Such identification is not intended to imply recommendation or endorsement by NIST, nor is it intended to imply that the entities, materials, or equipment are necessarily the best available for the purpose.

There may be references in this publication to other publications currently under development by NIST in accordance with its assigned statutory responsibilities. The information in this publication, including concepts and methodologies, may be used by Federal agencies even before the completion of such companion publications. Thus, until each publication is completed, current requirements, guidelines, and procedures, where they exist, remain operative. For planning and transition purposes, Federal agencies may wish to closely follow the development of these new publications by NIST.

Organizations are encouraged to review all draft publications during public comment periods and provide feedback to NIST. All NIST publications, other than the ones noted above, are available at http://csrc.nist.gov/publications.

NISTIR 7916

Abstract

Proceedings of the Cybersecurity in Cyber-Physical Workshop, April 23 – 24, 2012, complete with abstracts and slides from presenters. Some of the cyber-physical systems covered during the first day of the workshop included networked automotive vehicles, networked medical devices, semi-conductor manufacturing, and cyber-physical testbeds. Day two of the workshop covered the electric smart grid. Dr. Farnham Jahanian, NSF, was the keynote speaker on day one.

Keywords

CPS; cyber-physical systems; cybersecurity; networked automotive vehicles; networked medical devices; semi-conductor manufacturing

Acknowledgements

The editor would like to thank Suzanne Lightman, Kevin Stine (both NIST), and Mark Enstrom (Neustar) for being reviewers for this document.

Table of Contents

1. Introduction

Cyber-Physical Systems (CPS) are hybrid networked cyber and engineered physical elements co-designed to create adaptive and predictive systems for enhanced performance.[1,2] These smart systems present a key opportunity to create a competitive advantage for U.S. industrial innovation and to improve the performance and reliability of new and existing systems. From smart manufacturing and the electric smart grid, to smart structures and transportation systems, CPS will pervasively impact the economy and society.

Cybersecurity is a critical cross-cutting discipline that provides confidence that cyber-physical systems, their information, and supporting communications and information infrastructures are adequately safeguarded. CPS are increasingly being utilized in critical infrastructures and other settings. However, CPS have many unique characteristics, including the need for real-time response and extremely high availability, predictability, and reliability, which impacts cybersecurity decisions.

NIST is currently working in several CPS areas, including the electric smart grid, smart manufacturing, smart buildings, and networked automobiles. This work is being led by the Engineering Laboratory (EL), but also includes the Physical Measurement Laboratory (PML) and the Information Technology Laboratory (ITL). ITL has a number of key areas of expertise that are important to the evolution of CPS–interoperability, usability, reliability, and security. Since 2009, NIST has been very active in the area of the smart grid. ITL has been very active, providing leadership and expertise in a number of relevant areas, including communication networks, timing, and cybersecurity. ITL's Computer Security Division (CSD) has provided leadership and expertise to the Smart Grid Interoperability Panel's Cyber Security Working Group. Leveraging our broad expertise in relevant areas, we are now looking at the broader landscape of cyber-physical systems, and how cybersecurity fits into that landscape.

CSD hosted a two-day workshop to explore CPS cybersecurity needs, with a focus on research results and real-world deployment experiences. On the first day, speakers addressed CPS across multiple sectors of industry (e.g., automotive, healthcare, semi-conductor manufacturing). The second day focused on cyber security needs of CPS in the electric smart grid.

This document provides abstracts and corresponding slides from the plenary presentations at the workshop.[3]

---

[1] For more information on this definition, please see George Arnold's slides from the workshop.

[2] Performance metrics include safety and security, reliability, agility and stability, efficiency and sustainability.

[3] The website for the event is at http://www.nist.gov/itl/csd/cyberphysical-workshop.cfm, and the agenda is available at http://csrc.nist.gov/news_events/cps-workshop/cps-workshop-agenda_04-03-2012.pdf. The agenda document has links to electronic copies of the abstracts and slides.

2. Opening Remarks

George W. Arnold, DESc
National Coordinator, Smart Grid Interoperability
Director, Cyber-Physical Systems
Engineering Laboratory
NIST

Dr. Arnold joined the National Institute of Standards and Technology (NIST) in September 2006 as Deputy Director, Technology Services, after a 33-year career in the telecommunications and information technology industry. He was appointed National Coordinator for Smart Grid Interoperability at in April 2009. He has been responsible for leading the development of standards underpinning the nation's Smart Grid. In October 2011, Dr. Arnold added an additional role as Director of Cyber Physical Systems within NIST's Engineering Laboratory (EL). Anticipating and meeting the measurement science and standards needs for technology-intensive manufacturing construction, and cyber-physical systems in ways that enhance economic prosperity and improve the quality of life , EL promotes U.S. innovation and industrial competitiveness in areas of critical national priority.

Dr. Arnold served as Chairman of the Board of the American National Standards Institute (ANSI), a private, non-profit organization that coordinates the U.S. voluntary standardization and conformity assessment system, from 2003 to 2005. He served as President of the IEEE Standards Association in 2007-2008 and as Vice President-Policy for the International Organization for Standardization (ISO) from 2006-2009 where he was responsible for guiding ISO's strategic plan.

Dr. Arnold previously served as a Vice-President at Lucent Technologies Bell Laboratories where he directed the company's global standards efforts. His organization played a leading role in the development of international standards for Intelligent Networks and IP-based Next Generation Networks. In previous assignments at AT&T Bell Laboratories he had responsibilities in network planning, systems engineering, and application of information technology to automate operations and maintenance of the nationwide telecommunications network.

Dr. Arnold received a Doctor of Engineering Science degree in Electrical Engineering and Computer Science from Columbia University in 1978. He is a Fellow of the IEEE.

Opening Remarks

Cybersecurity in Cyber-Physical Systems Workshop
hosted by
NIST Information Technology Laboratory
April 23-24, 2012

George W. Arnold, Eng.Sc.D.
Director, Smart Grid and Cyber-Physical Systems Program Office
Engineering Laboratory
National Institute of Standards and Technology
U.S. Department of Commerce

cyber-physical systems

NIST At A Glance

Gaithersburg, MD

Boulder, CO

- NIST Research Laboratories
- Manufacturing Extension Partnership
- Baldrige Performance Excellence Award
- Technology Innovation Program

~ 2,900 employees

~ 2,600 associates and facility users

~ 1,600 field staff in partner organizations

~ 400 NIST staff serving on 1,000 national and international standards committees

cyber physical systems

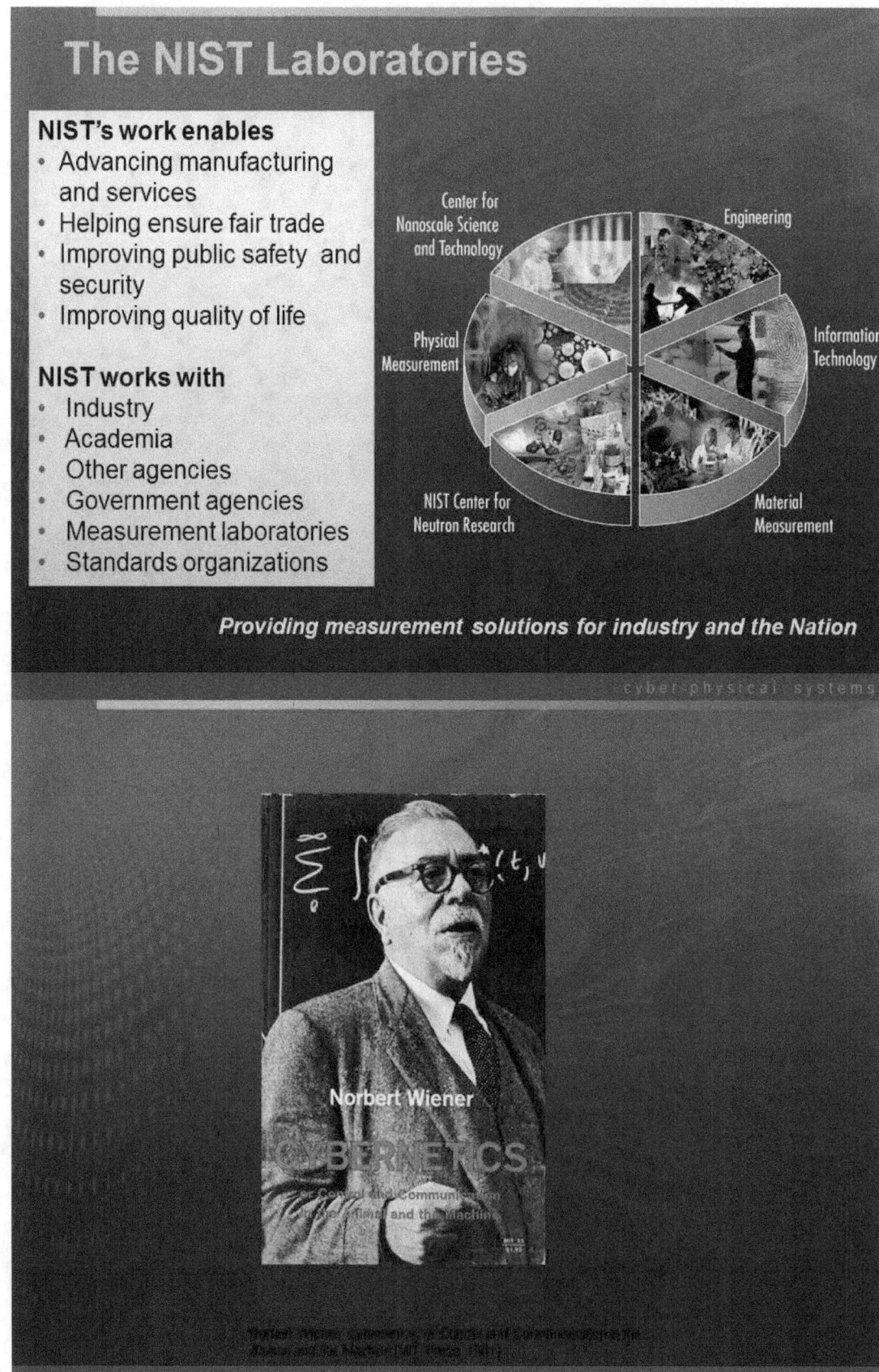

Smart Grid: An Example of a CPS

NIST Smart Grid Reference Model

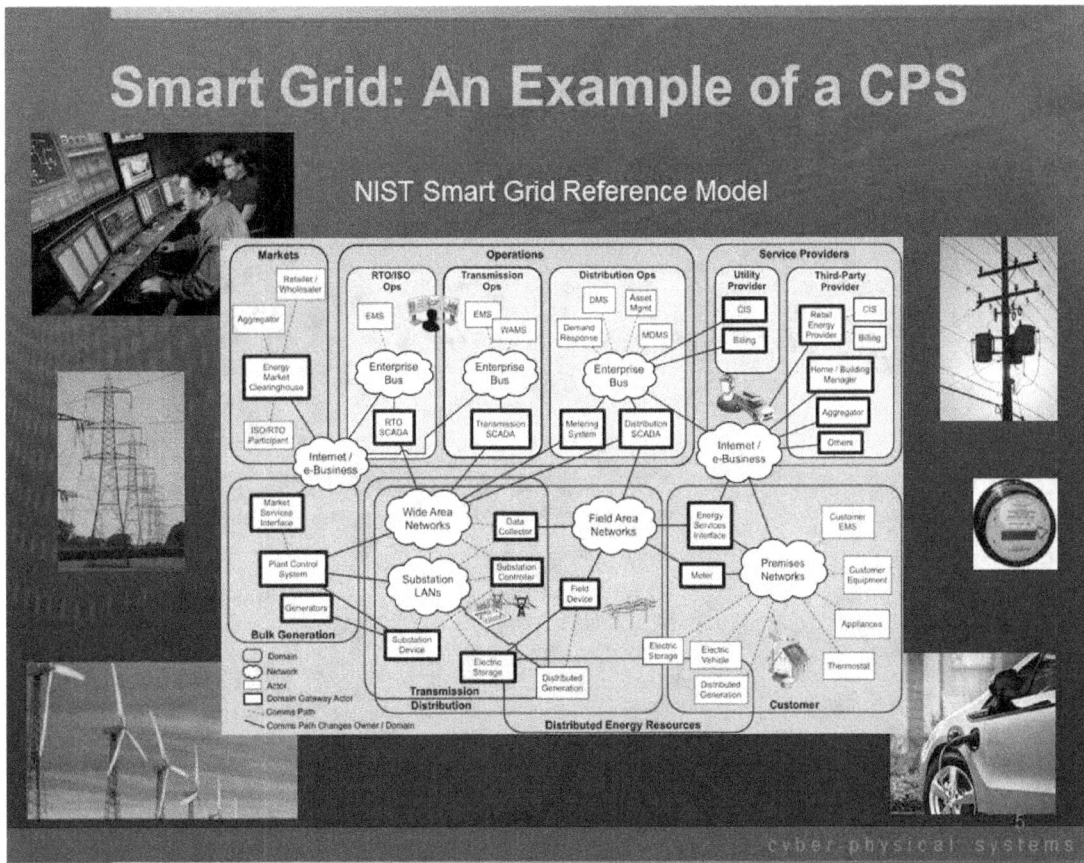

Smart Manufacturing: Another CPS Application

Smart Manufacturing refers to **manufacturing production systems at the equipment, factory, and enterprise levels** that integrate cyber and physical systems by combining:

- smart operating systems to **monitor, control, and optimize performance**
- systems engineering-based **architectures and standards,** and
- embedded and/or distributed **sensing, computing, communications, actuation, and control technologies**

to enable **innovative production**, **products**, and/or **systems of products** that enhance *economic and sustainability performance*

Definition of Cyber-Physical Systems

Function:

Cyber physical systems are hybrid networked cyber and engineered physical elements co-designed to create adaptive and predictive systems for enhanced performance[*]

Essential Characteristics:

- Co-design treats cyber, engineered, and human elements as integral components of a functional whole system to create synergy and enable desired, emergent properties

- Integration of deep physics-based and digital world models provides learning and predictive capabilities for decision support (e.g., diagnostics, prognostics) and autonomous function

- Systems engineering-based architectures and standards provide for modularity and composability for customization, systems of products, and complex or dynamic applications

- Reciprocal feedback loops between computational elements and distributed sensing/actuation and monitoring/control elements enables adaptive multi-objective performance

- Networked cyber components provide a basis for scalability, complexity management, and resilience

---

[*]Performance metrics include safety and security, reliability, agility and stability, efficiency and sustainability, privacy

cyber-physical systems

CPS Application Sectors and Benefits

Application Sectors:

- **Manufacturing** (includes smart production equipment, processes, automation, control, and networks; new product design)

- **Transportation** (includes intelligent vehicles and traffic control)

- **Infrastructure** (includes smart utility grids and smart buildings/structures)

- **Health Care** (includes body area networks and assistive systems)

- **Emergency Response** (includes detection and surveillance systems, communication networks, and emergency response equipment)

- **Warfighting** (includes soldier equipment systems, weapons systems and systems of systems, logistics systems)

Benefits:

- Improved **quality of life** and **economic security** through **innovative functions**, **production**, **products**, and/or **systems of products**

cyber-physical systems

NIST CPS Context

- Growing demands on NIST for standards associated with smart systems applications
  - Smart Buildings, Smart Grid and Infrastructure, Smart Manufacturing, Smart Health Care, Smart Transportation, ...
- NIST has responded with programs in individual domain areas
- Significant crosscutting technology gaps and fundamental research challenges exist
- Potential impact on manufacturing: Innovative new classes of manufactured products, systems of products, and production systems

cyber-physical systems

CPS Platform Technology Gaps and R&D Grand Challenges

- Platform Technology Gaps (Systems-Engineering Based Architectures and Standards)
  - Modularity and composability
  - Deep-physics and digital world model integration
  - Control, communications, and interoperability (adaptive and predictive; time synchronization)
  - Cyber-security
  - Scalability, complexity management, and resilience (integration with legacy systems)
  - Wireless sensing and actuation
  - Validation and verification; assurance and certification (software, controls, system)
- R&D Grand Challenges
  - Co-designing hybrid networked systems with integrated cyber, engineered, and human elements
  - Synthesizing and evolving complex, dynamic systems with predictable behavior (diagnostics, prognostics); anticipating emergent behaviors arising from interactions
  - Multi-scale, multi-physics modeling across discrete and continuous domains
  - Incorporating uncertainty and risk into reasoning and decision-making
  - Modeling and defining levels of autonomy and optimizing role of the human
  - Enabling education and workforce development; technology transfer

cyber-physical systems

NIST CPS Actions

- NIST CPS Working Group (EL, ITL, SCO, OLES; January 2011)
- Cooperative Agreement with UMD for CPS research (Kick-off December 2011)
  - Book assessing state-of-the-art
  - Market analysis to guide R&D investments
  - Platform-based architecture and standards framework
  - Fundamental research in modeling and synthesis
- Short Course for Executives delivered by world class industry and research leaders (January 19-20, 2012)
- R&D Needs Assessment Workshop: Foundations for Innovation in CPS (March 13-14, 2012)
- Performance Metrics for Intelligent Systems (PerMIS) Workshop – CPS Theme (March 20-22, 2012)
- Cyber-security for Cyber-Physical Systems Workshop (April 23-24)
- Planned CTO Roundtable (June 2012)

cyber-physical systems

Cybersecurity of CPS: New Challenges

- Need to address all the conventional aspects of cybersecurity, plus

- New issues and threats, e.g.

  - Complex software with non-deterministic behavior

  - Precise timing requirements

  - Cyber system as a threat vector for attack on the physical system rather than the object of attack

cyber-physical systems

3. Implantable Medical Devices – Cyber Risks and Mitigation Approaches

Sarbari Gupta

Electrosoft Services
Reston, Virginia USA
sarbari@electrosoft-inc.com

Abstract—Over the past decade, there has been an explosion in the deployment of implantable medical devices (IMDs) to facilitate the management and treatment of a wide variety of human health conditions. While functionality and patient safety requirements have driven new generation IMDs to be increasingly accessible through wireless communication channels, these changes cause significant concern in terms of increased risk from cyber threats whether malicious or unintentional. This paper investigates the risks associated with such devices from the cyber environment and proposes approaches to support decisions regarding the integration of adequate security and privacy measures to mitigate these risks.

Keywords-medical devices; security, privacy, cyber, risks

I. Introduction

Deployment rates for implantable medical devices (IMDs) have skyrocketed over the past decade. Devices such as pacemakers, cardiac defibrillators, heart monitors, cochlear implants, insulin pumps, infusion pumps and other similar devices are routinely used to monitor and treat a plethora of medical conditions. These IMDs have been increasingly accessible through wireless channels to support functions such as emergency extraction of patient health history, remote monitoring of health status, firmware updates and local as well as remote therapy reprogramming. As with all things connected to the cyber world, there are known and unknown threats lurking that threaten the reliability and safety of these devices as well the privacy of patients who depends on them.

II. Regulation of Implantable Medical Devices

Within the United States, the Food and Drug Administration (FDA) regulates the manufacturers, importers and resellers of these devices through the Center for Devices and Radiological Health (CDRH). A review of a sampling of FDA testing guidance (e.g. for implantable cardiac pacemakers) reveals that the "tests are designed to reasonably assure safe and effective functioning of the pacemaker in the patient, according to written specifications of performance, and its survival under expected environmental conditions in the body and during storage, shipping and handling" [1]. FDA testing guidelines do not appear to address the resistance and resilience of these devices in the face of cyber attacks.

III. Review of Recent Research

Halperin et al have shown that a recent (2003) model of implantable cardioverter defibrillator (ICD), designed to communicate wirelessly with an external programmer in the 175 kHz frequency range, is vulnerable to several radio-based attacks that threaten patient safety and privacy [2]. Several other research papers have pointed out similar vulnerabilities to cyber threats and possible mitigation mechanisms [3, 4, 5, 6].

IV. Security Analysis of IMDs

IMDs are tiny computing platforms that run firmware in extremely power constrained environments. They offer data storage for static data (such as device information), relatively static data (such as patient identification, medical condition, therapy configuration) and dynamic data (such as recent patient readings and audit logs). IMDs offer wireless access for read and write operations to the data on the IMD (including the firmware) to a variety of stakeholders and roles.

A. Threats, Vulnerabilities and Risks

Some of the threats to wireless IMDs include device reprogramming, data extraction, data tampering, repeated access attempts and data flooding. Vulnerabilities include unsecured communication channels, inadequate authentication and access control, weak audit mechanisms and meager storage. The resulting risks include patient safety compromise resulting from firmware malfunction or therapy misconfiguration, device unavailability due to battery power depletion, patient privacy loss due to data leakage to unauthorized parties, and inappropriate medical follow-up due to tampering of patient readings. While some cyber threats may be unintentional, various motivations exist for deliberate cyber attacks, such as patient information gathering, negative impact to patient health status, ego satisfaction of the attacker, as well as gaining competitive advantage over another vendor through negative press.

B. Impact of Security Compromise

Identification of the various data types within an IMD is an essential step in analyzing the security and privacy risks of such devices. Possible data types include firmware (though technically not "data"), device identification data, patient identification and health condition data, therapy configuration data, patient readings, audit log data, and other data.

Following identification of the different data types within the IMD, it is useful to conduct a security categorization using the approach described in FIPS 199 [7]. For each data type, the security analyst asks the question: "What is the impact (High, Moderate or Low) of a compromise to the confidentiality, integrity and availability of this type of data?" It is useful to collect the results of this analysis in a table format.

C. Authentication and Access Control Mechanisms

For each type of data identified, the authentication and access control mechanisms applicable for extracting or updating the data type need to be reviewed to determine adequacy of the protection mechanisms while balancing the needs of patient safety in emergency situations and the utility of the IMD within a patient's environment. This is a non-trivial exercise since the security and privacy requirements for IMDs frequently conflict with the requirements stemming from emergency access to patient data and device utility in hospital and home settings. Creative approaches may be devised to decouple data essential for patient safety in emergency conditions from patient personally identifiable information/data to allow different authentication and access control mechanisms to apply to each group of data. Alternately, identifying different modes of operation (such as home health setting versus open environment versus emergency situation) to allow the IMD to apply different authentication and access control mechanisms in different modes.

D. Cryptographic Techniques

Cryptographic techniques are potentially very useful to improve the security and privacy properties of IMDs through stronger authentication protocols and (confidentiality and integrity) protected communications over wireless channels. However, since IMDs operate in very constrained environments (such as device size, cost, and power availability,) traditional cryptographic techniques and protocols may be inappropriate. More compatible cryptographic suites and protocols need to be devised for use on IMDs and applied in a very selective manner to optimize the security protection from these power intensive operations. The body of research conducted for cryptography for sensor networks are directly applicable [8] to applying cryptographic techniques to IMDs.

E. Audit Mechanisms

Audit logs are essential for tracking patient history and IMD behavior over a period of time. The audit records provide information needed for adequate patient care as well as updates to patient therapy delivered through the IMD. Given the limited storage capabilities of IMDs, it is possible to overflow the audit logs through certain types of attacks on the IMD. Creative techniques for selective overwriting of audit records based on significance of each type of audit record may be useful. Alert mechanisms when audit log storage space nears depletion may also be useful for alerting the patient or the remote monitoring facility so that appropriate steps can be taken in a timely manner prior to audit space exhaustion.

V. Summary and Next Steps

Implantable medical devices pose a number of security and privacy risks even while providing essential medical support functions such as patient monitoring and treatment delivery. With the proliferation of IMDs of various types, it is essential to understand the risks from cyber threats, and integrate sufficient protections and controls to balance patient safety and device utility with security and privacy risks.

Some of the possible next steps in this area include (i) applying risk assessment methods to better understand the threat model and risks applicable to each type of IMD, (ii) performing security categorization analyses to various data types to guide optimal grouping of data to better protect each data group and apply appropriate cryptographic techniques when appropriate, (iii) development of guidelines for development, delivery, configuration, and monitoring of IMDs, and (iv) targeted regulation of IMDs by the FDA CDRH (in the United States) to improve protection against cyber risks. .

References

[1] "Implantable Pacemaker Testing Guidance," found at http://www.fda.gov/downloads/MedicalDevices/DeviceRegulationandGuidance/GuidanceDocuments/UCM081382.pdf.

[2] D. Halperin, et al, "Pacemakers and Implantable Cardiac Defibrillators: Software Radio Attacks and Zero-Power Defenses," Proceedings of the 2008 IEEE Symposium on Security and Privacy, Oakland, CA, 2008.

[3] D. Halperin et al, "Security and Privacy for Implantable Medical Devices," in Pervasive Computing, Vol. 7, No. 1, January–March 2008.

[4] S. Capkun, "On Secure Access to Medical Implants," Workshop on Security and Privacy in Implantable Medical Devices, Lausanne, Switzerland, April, 2011.

[5] S. Cherukuri, K. Venkatasubramanian, and S. Gupta, "BioSec: A Biometric Based Approach for Securing Communication in Wireless Networks of Biosensors Implanted in the Human Body," Proc. Int'l Conf. Parallel Processing (ICPP)Workshops, IEEE CS Press, 2003, pp. 432–439.

[6] T. Denning, et al "Patients, pacemakers, and implantable defibrillators: human values and security for wireless implantable medical devices," Proceedings of the 28th international conference on Human factors in computing systems, ACM New York, NY, USA, 2010, pp 917-926.

[7] National Institute of Standards and Technology "FIPS Pub 199: Standards for Security Categorization of Federal Information and Information Systems," FEDERAL INFORMATION PROCESSING STANDARDS PUBLICATION, February 2004.

[8] S. Fischer and M. Zitterbart, "Security in Sensor Networks," Information Technology: Vol. 52, No. 6, 2010, pp. 311-312.

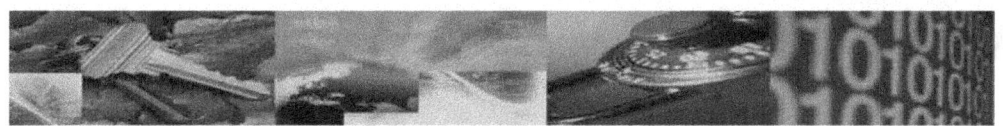

Implantable Medical Devices – Cyber Risks and Mitigation Approaches

NIST Cyber Physical Systems Workshop
April 23-24, 2012

Dr. Sarbari Gupta, CISSP, CISA
sarbari@electrosoft-inc.com; 703-437-9451 ext 12

Electrosoft

- **Overview of IMDs**
- **Security Threats, Vulnerabilities and Risks**
- **Risk-Based Mitigation Approach**
- **Summary**
- **References**

Electrosoft

What is an IMD?

- **Implantable Medical Device (IMD)**
  - **Tiny computing platform with firmware**
  - **Runs on small batteries**
  - **Programmable**
  - **Implanted in human body**
  - **Monitors health status**
  - **Delivers medical therapy**

Electrosoft

Page 3

IMD Examples

- **Pacemakers**
- **Implantable Cardiac Defibrillators (ICD)**
- **Cochlear Implants**
- **Insulin Pumps**
- **Neurostimulators**

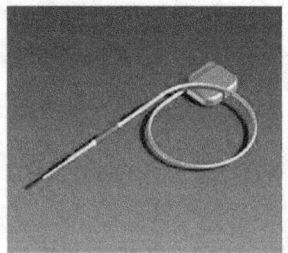

Electrosoft

Page 4

Wireless Implantable Medical Devices

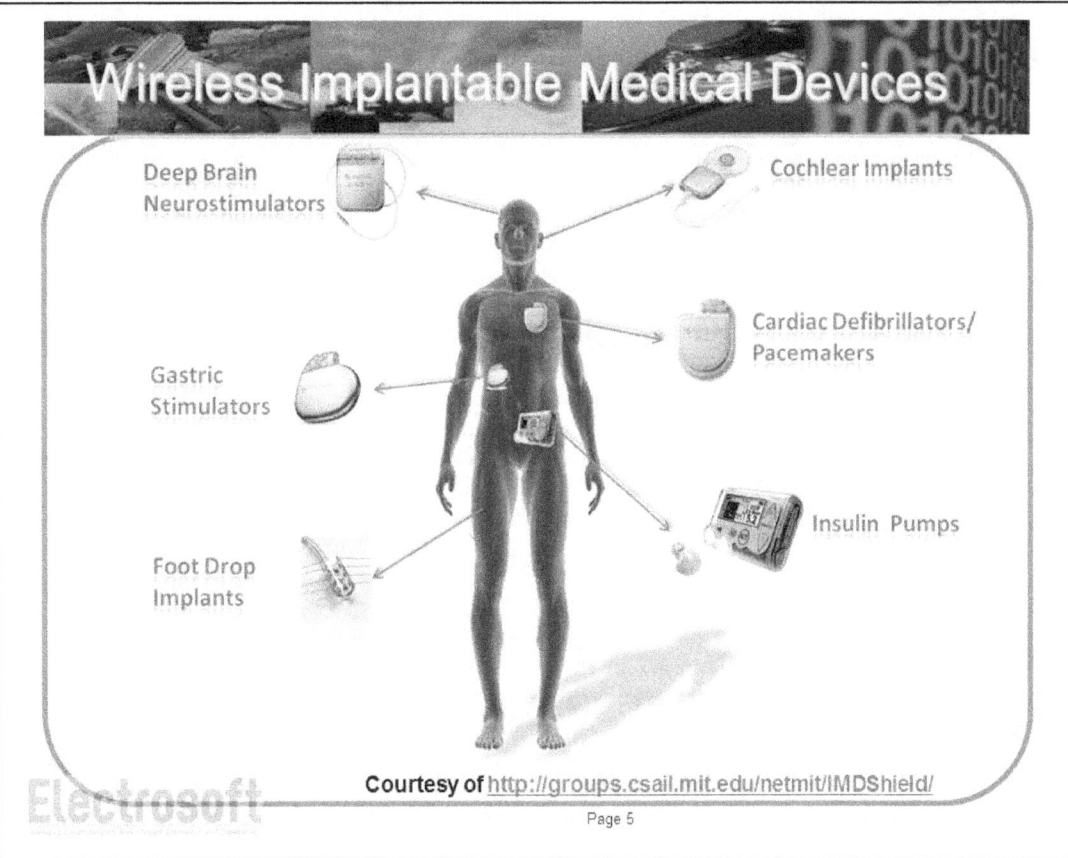

- Deep Brain Neurostimulators
- Cochlear Implants
- Gastric Stimulators
- Cardiac Defibrillators/ Pacemakers
- Foot Drop Implants
- Insulin Pumps

Courtesy of http://groups.csail.mit.edu/netmit/IMDShield/

Page 5

Pacemaker

- **Consists of battery, computerized generator, and wires with sensors at tips (pacing leads)**
  - **Wires connect generator to the heart**

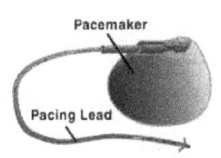
Pacemaker
Pacing Lead

- **Records heart's electrical activity and rhythm**
  - **Recordings used to adjust pacemaker therapy**

- **On abnormal heart rhythm**
  - **Generator sends electrical pulses to heart**

- **Can monitor blood temperature, breathing etc.**
  - **Can adjust heart rate to changes in your activity**

- **Wireless communication with Programmer**
  - **Read battery status and heart rhythms**
  - **Send instructions to change therapy**

Page 6

13

Wireless Insulin Pump

- Supports blood sugar monitoring & insulin delivery

- Wireless integration of Monitor and Pump

- Pump pre-set with user-specific information

- Monitor transmits glucose value to pump via wireless

- Pump calculates and delivers proper insulin dosage

- Pump "remembers" dosage history

- PC "dongle" can connect to Pump to read data or update settings

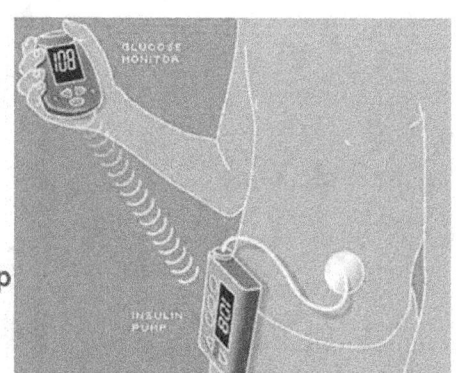

Medtronic Paradigm 512 Insulin Pump with Wireless Blood Sugar Meter

Page 7

Cochlear Implants

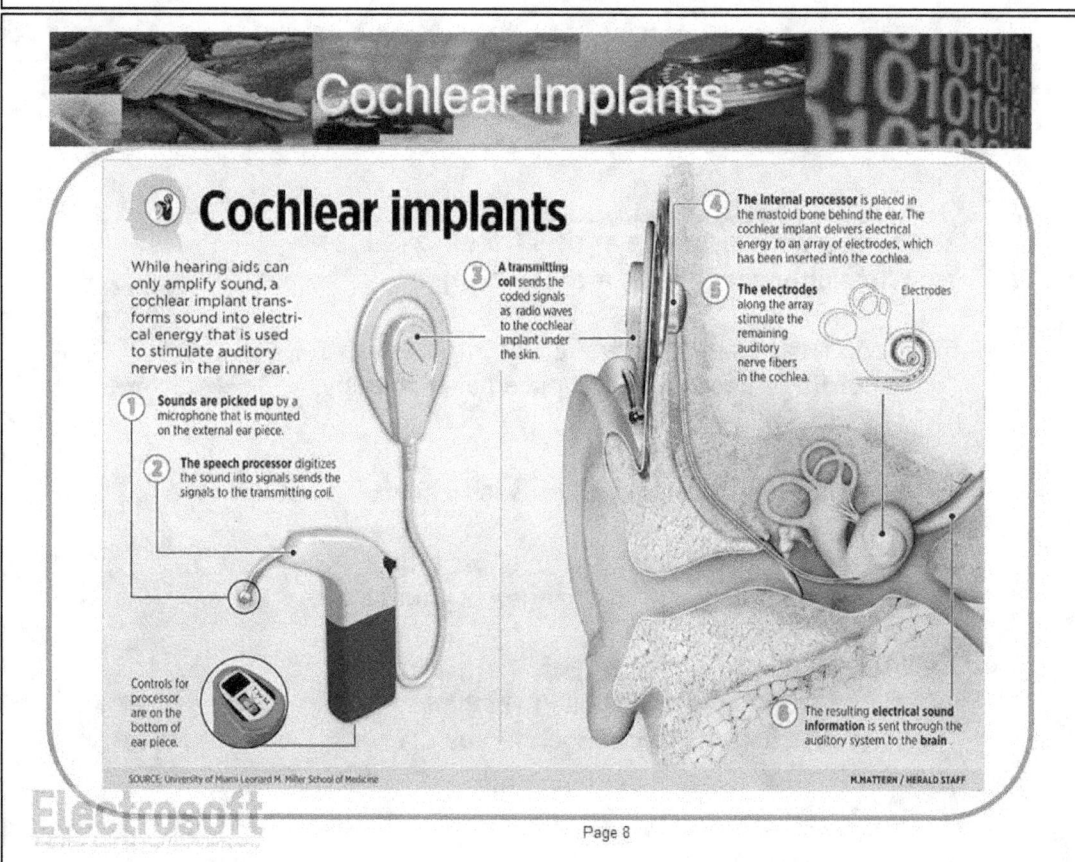

Page 8

IMD Data

- **IMD holds various Data Types**
  - **Static Data**
    - Device make
    - Model #
  - **Semi-static Data**
    - Physician & Health Center Identification
    - Patient Name and DOB
    - Medical condition
    - Therapy configuration
  - **Dynamic Data**
    - Patient health status history
    - Therapy and dosage history
    - Audit logs

Electrosoft

IMD Accessibility

- **"Programmer" Device communicates with IMD**
  - **Through wireless channels**
  - **Using radio frequency transmission**

- **PC communicates with IMD**
  - **Through USB-port "dongles" using radio frequencies**
  - **PC may also be connected to Internet**

- **IMD functions accessed remotely**
  - **Read data on health status & therapy history**
  - **Emergency extraction of patient health history**
  - **Emergency reset of IMD configuration**
  - **Therapy programming/reprogramming**
  - **Firmware updates**

Electrosoft

Regulation of IMDs

- In US, IMDs are regulated by
  - Food and Drug Administration (FDA) Center for Devices and Radiological Health (CDRH)

- Testing focus
  - Safe and effective functioning
  - Different environmental conditions

- Absence of focus
  - Resistance/Resilience to cyber attacks

Electrosoft

Page 11

Are IMDs Vulnerable?

- A resounding YES!

- Current devices are engineered without considering threat of a potential hacker

- Current methods to prevent unauthorized access to IMDs include
  - Use of proprietary protocols
  - Controlled access to "Programmers" devices
  - Essentially, *security by obscurity!*

Electrosoft

Page 12

16

Black Hat security conference – Aug 2011

- "Security researcher Jerome Radcliffe has detailed how our use of SCADA insulin pumps, pacemakers, and implanted defibrillators could lead to **untraceable, lethal attacks from half a mile away**"

- "He managed to **intercept the wireless control signals, reverse them, inject some fake data**, and then send it back to the [insulin] pump."

- "He could increase the amount of insulin injected by the pump, or reduce it"

http://www.extremetech.com/extreme/92054-black-hat-hacker-details-wireless-attack-on-insulin-pumps

IEEE Symposium on Security and Privacy – 2008

- Halperin et al, "Pacemakers and Implantable Cardiac Defibrillators: Software Radio Attacks and Zero-Power Defenses"

- "… an implantable cardioverter defibrillator (1) is potentially **susceptible to malicious attacks that violate the privacy** of patient information and medical telemetry, and (2) may experience **malicious alteration to the integrity of information or state**, including patient data and therapy settings for when and how shocks are administered."

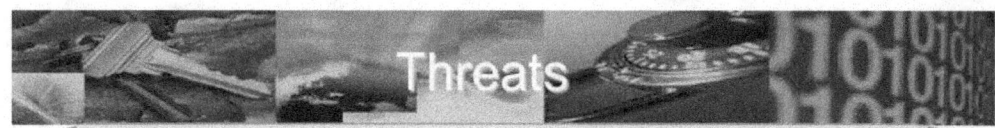

Threats

- Patient Data Extraction
- Patient Data Tampering
- Device Re-programming
- Repeated Access Attempts
- Device Shut-Off
- Therapy Update
- Malicious Inputs
- Data Flooding

Page 15

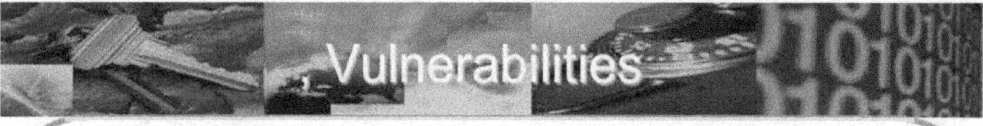

Vulnerabilities

- Unsecured Communication Channels
- Inadequate Authentication Mechanisms
- Inadequate Access Controls
- Software Vulnerabilities
- Weak Audit Mechanisms
- Meager Storage
- Insufficient Alerts

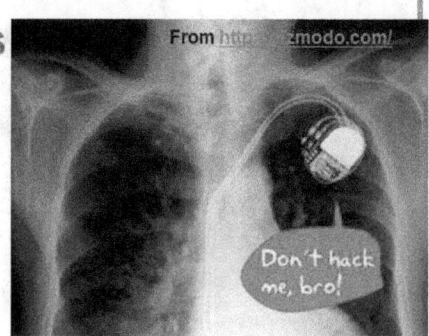

Page 16

18

Risks

- Patient Health Safety
  - Firmware Malfunction
  - Malicious Therapy Update
  - Malicious Inputs to Device
- Patient Privacy Loss
  - Data Leakage from Device
- Inappropriate Medical Follow-up
  - Tampering of Patient Readings
- Device Unavailability
  - Battery Power Depletion
  - Device Flooding

Risk-Based Mitigation Approach

- Develop IMD Security Impact Matrix
- Develop IMD Access Requirements Matrix
- Select Appropriate Security Mechanisms
- Tailor Security Mechanisms
  - Accommodate IMD Environment Constraints
  - Add Compensating Mechanisms (as needed)

FIPS 199-based Impact Analysis

- Identify IMD Data Types
  - **E.g., Firmware, Device Identification, Patient Identification, Provider Identification, Health Condition, Therapy Configuration, Patient Readings, Audit Logs**

- Identify IMD Health Delivery Commands
  - **E.g., Emergency reset**

- Analyze Impact of Compromise
  - **For each Data Type, estimate impact**
    - o Loss of Confidentiality, Integrity and Availability
  - **For each Command Type, estimate impact**
    - o Loss of Availability
  - **Assign Impact as [LOW, MODERATE, HIGH]**

- Tabulate in IMD Security Impact Matrix

Page 19

IMD Security Impact Matrix (IMD-SIM)

Security Function / Data, Command	Emergency Reset Command	Patient ID Data	Therapy Data	Patient Heath Data
Confidentiality	N/A	MOD	LOW	MOD
Integrity	N/A	MOD	HIGH	HIGH
Availability	HIGH	LOW	MOD	MOD

Page 20

20

Determine IMD Access Requirements

- **Develop Matrix**
  - **By Data Type and Health Delivery Command**
  - **By Role of Individual Accessing IMD and**
    - By Access Channels (e.g., wired, wireless)
- **Add Required Access Privileges**
  - **Per Basic IMD Functionality**
  - **By Need for Emergency Access**
  - **By Utility and Quality of Life Factors**
- **Tabulate as IMD Access Requirements Matrix (IMD-ARM)**

Electrosoft

Page 21

IMD Access Requirements Matrix (IMD-ARM)

ROLE-CHANNEL / Command, Data	Emergency Reset Cmd	Patient ID Data	Therapy Data	Patient Heath Data
Patient-Wireless				
Prescribing Physician-Wired		Read Write	Read Write	Read
Maintenance Physician-Wireless		Read	Read	Read
Emergency Tech-Wireless	Invoke			

Electrosoft

Page 22

21

Select Needed Security Mechanisms

- Overlay IMD-IAM and IMD-ARM
- Select Security Mechanisms to Protect IMD Data/Commands
  - **Channel Protection Mechanisms**
    - o Crypto-protected channel
    - o None (Proprietary Protocols)
  - **Authentication Mechanisms**
    - o Password
    - o Device-to-device handshake
    - o Cryptographic authentication
  - **Audit Mechanisms**
    - o Auditable Events
    - o Management of Audit Space Depletion
  - **Alert/Alarm Mechanisms**
    - o Audible Alarms
    - o Automatic Device Reset to Safe Mode

Page 23

Tailor Security Mechanisms

- **IMDs subject to many constraints**
  - **Device Size**
  - **Cost**
  - **Power**
  - **Computational Capability**
  - **Storage**
- **Adjust security mechanisms to accommodate constraints**
  - **E.g., Add Alarm if authentication can't be strengthened for certain Data Types**

Page 24

Special Challenges in Securing IMDs

- Battery and Power Limitations
  - Power usage must be minimized to extend battery life
  - Battery depletion has devastating health consequences

- Use of Cryptographic Techniques
  - Highly Constrained Environment (cost, power, storage)
  - Compatible Crypto Suites/Protocols Needed
    - Crypto for Sensor Networks

- Audit Mechanisms
  - Limited Storage Area on Device
    - Attacks may generate deluge of audit entries
  - Managing Audit Space Depletion
    - Selective Overwriting; Alarms (Audible or to Remote Monitor)

Electrosoft

Page 25

Summary – IMDs and Security

- IMDs – Essential in Current Healthcare Environment
- Wireless Access
  - Promotes Usability and Utility
  - Poses Significant Security and Privacy Concerns
- Risk-based Mitigation Approach
  - Determine Security Impact for Data Types
  - Implement Adequate Security Mechanisms
  - Balance Security/Privacy with Safety/Usability
- Further Work
  - Models for IMD security and privacy
  - Crypto-suites for IMD environments

Electrosoft

Page 26

23

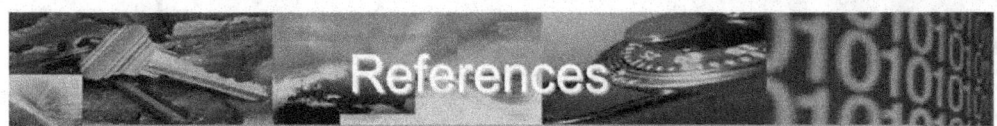

References

- "Implantable Pacemaker Testing Guidance," http://www.fda.gov/downloads/MedicalDevices/DeviceRegulationandGuidance/GuidanceDocuments/UCM081382.pdf.
- D. Halperin, et al, "Pacemakers and Implantable Cardiac Defibrillators: Software Radio Attacks and Zero-Power Defenses," Proceedings of the 2008 IEEE Symposium on Security and Privacy, Oakland, CA, 2008.
- D. Halperin et al, "Security and Privacy for Implantable Medical Devices," in Pervasive Computing, Vol. 7, No. 1, January–March 2008.
- S. Capkun, "On Secure Access to Medical Implants," Workshop on Security and Privacy in Implantable Medical Devices, Lausanne, Switzerland, April, 2011.
- S. Cherukuri, K. Venkatasubramanian, and S. Gupta, "BioSec: A Biometric Based Approach for Securing Communication in Wireless Networks of Biosensors Implanted in the Human Body," Proc. Int'l Conf. Parallel Processing (ICPP)Workshops, IEEE CS Press, 2003, pp. 432–439.
- T. Denning, et al "Patients, pacemakers, and implantable defibrillators: human values and security for wireless implantable medical devices," Proceedings of the 28th international conference on Human factors in computing systems, ACM New York, NY, USA, 2010, pp 917-926.
- National Institute of Standards and Technology "FIPS Pub 199: Standards for Security Categorization of Federal Information and Information Systems," FEDERAL INFORMATION PROCESSING STANDARDS PUBLICATION, February 2004.
- S. Fischer and M. Zitterbart, "Security in Sensor Networks," Information Technology: Vol. 52, No. 6, 2010, pp. 311-312.

Electrosoft

Page 27

Questions and Contact Information

- ## Dr. Sarbari Gupta – Electrosoft
  - **Email:** sarbari@electrosoft-inc.com
  - **Phone: 703-437-9451 ext 12**
  - **LinkedIn:** http://www.linkedin.com/profile/view?id=8759633

Electrosoft

Page 28

4. Safety-Critical Automotive and Industrial Data Security

André Weimerskirch

ESCRYPT Inc.

Ann Arbor, MI, USA

andre.weimerskirch@escrypt.com

I. Introduction

Automotive and industrial data security is researched for almost a decade now and the author started doing research and working in this area in 2003. Recent attacks impressively demonstrated weaknesses that were anticipated for a while now. In the area of automotive data security, a research team of the University of Washington and University of California, San Diego, was able to hack into a modern vehicle and control the vehicle [2][4]. The team mounted attacks via external interfaces, such as Bluetooth and cellular connection, and internal interfaces, such as USB flash drive and CD. The research team was then able to replace the firmware of safety critical components and was thus potentially able to crash the vehicle. Bailey presented an attack at the Black Hat congress to undermine the remote unlock and remote start mechanism of a car via smartphone [1]. Similar threads also exist in less researched areas, such as automatic mining, industry production robots, and construction site machines. Even in very remote areas similar concerns arise. For instance, advanced fire alarm systems (e.g. for an office building) are controlled by an embedded computing system and the compromise of such a system might be fatal.

Data security and privacy is well understood for regular Internet systems, consisting of PCs, servers, network equipment, etc. However, even there no proper security strategies are in place for the majority of systems, as shown by the daily news about compromised financial institutions, government organizations, and critical infrastructure components. The situation is very different in automotive and industrial security systems. Unfortunately, this difference is not well understood and very often leads to poor security design and security weaknesses in the first place. Fortunately, no actual attack was ever reported to automotive and industrial systems. However, we believe it is only a matter of time until the knowledge becomes widespread and attacks will be mounted. We believe that security in the automotive area is most researched and understood, and that the results can be applied to further industrial security systems such as machines, industry robots, fire alarm control systems, etc. Therefore the remainder of this article will often make references to automotive security systems.

II. Background

The threat model for safety-critical automotive and industrial systems is quite different to traditional network systems. Comfort and remote maintenance features are connected to safety critical systems. For instance, in a passenger vehicle there is a physical network connection, typically via CAN bus, between the infotainment system (that in turn might be connected to Bluetooth, Wi-Fi, and cellular data connection) and the safety critical powertrain components. Especially during the last few years, there is an increased desire to provide communication features due to raised consumer expectations. Consumers expect a vehicle with infotainment systems that resembles modern smart-phone comfort features and that provide Internet connections. Industry robot and machine owners expect remote control and maintenance features. At the same time, cost pressure does not allow implementing failure-safe security mechanisms (e.g. by using two physically separated communication bus systems within a vehicle, with redundant components that are connected to both bus systems). The threat model for safety-critical automotive and industrial systems is summarized in the following:

- **Assumptions and limitations**: automotive and industrial systems often provide physical access to the devices. However, these systems do not provide a permanent Internet connection and it is often not possible to regularly update software, as we are used to from the PC world. In fact, for today's passenger vehicles software updates are only performed upon customer's demand or in case of noticeable malfunction.

- **Attacker motivation**: as of today, there are no known attacks, mainly due to the significant effort required to mount attacks and due to the missing motivation. In particular, there is no financial motivation. The more business models are introduced, e.g. subscription services for the infotainment platform, and the more motivation there is for attackers to undermine the system. Attackers might then extend their attacks due to curiosity, or they might accidentally uncover safety critical attacks. Another potential group of attacker belongs to the curious hacker on the hunt for spectacular hacks.

- **Attack targets:** potential targets are the safety critical components, the remote maintenance feature, and undermining financial business models. Attackers might target competitors to deactivate machines in a construction site, and attackers might offer their services as an illegal business to interested parties. A further attack target is the extraction of information, e.g. from the devices of a competitor, in order to gain confidential and privacy-sensitive information.

- **Likelihood of attack:** today the effort to mount an attack in terms of knowledge and financial resources is significant, and there are easier and less costly ways to harm vehicle passengers. However, once the knowledge becomes widespread, and once attacks can be mounted very easily, the likelihood of attacks will increase.

- **Impact and risk of attack:** the impact of attacks is significant, thus leading to a high risk level. A successful attack can potentially harm people.

III. Countermeasures

Currently there are no legal requirements or guidelines available to the manufacturers of such systems. There is also no security standardization available. However, there are several research projects that will provide approaches to counter the described attacks. We believe that security in such systems needs to be approached by considering the following layers:

1. **Applications and operating system:** applications shall be implemented using current state-of-the-art knowledge and proper processes. For instance, there shall be no software modules included that is not actually needed (often used when legacy systems or open source software is used).

2. **Virtualization, hyper threading & microkernel:** We believe that it is impossible to implement applications and a full-blown operating system without security weaknesses that will be discovered over the life-span of the device. Therefore we suggest the use of virtualization and microkernel technology. The microkernel is a relatively small kernel (around 10,000 lines of code) that only provides the essential kernel features. Since the kernel is fairly small in terms of source code, it can be assumed that there are no significant security weaknesses in the microkernel. The actual operating system and applications visible to the user are executed in a compartment. If a compartment is hacked, the attack is limited to the confinement of the compartment. The European Union funded OVERSEE project [5].

3. **Secure hardware:** attacks can potentially endanger safety of life and therefore we suggest introducing a final security barrier at the hardware layer. Such a solution must be cost efficient due to the cost pressure. The European Union funded EVITA project [3] considers secure computing platforms for automotive systems. Furthermore, the equivalent of firewalls or gateways can be introduced to control traffic between the comfort and maintenance components, and the safety critical components.

IV. Outlook

The full presentation will provide an overview of today's attacks and will detail the attacker model. Special consideration will be given to available countermeasures and the most interesting research projects will be described. Finally, suggestions for improvements will be made. These might include security certifications for safety critical systems, such as Common Criteria and FIPS 140-2 security certifications, and it might be wise to setup a CERT for safety critical automotive and industrial systems.

References

[1] Don Bailey, "War Texting: Identifying and Interacting with Devices on the Telephone Network", Blackhat USA, 2011.

[2] Stephen Checkoway, Damon McCoy, Brian Kantor, Danny Anderson, Hovav Shacham, Stefan Savage, Karl Koscher, Alexei Czeskis, Franziska Roesner, Tadayoshi Kohno, "Comprehensive Experimental Analysis of Automotive Attack Surfaces". USENIX Security, August 10–12, 2011.

[3] EVITA, "E-safety vehicle intrusion protected applications", http://www.evita-project.org

[4] Karl Koscher, Alexei Czeskis, Franziska Roesner, Shwetak Patel, Tadayoshi Kohno, Stephen Checkoway, Damon McCoy, Brian Kantor, Danny Anderson, Hovav Shacham, Stefan Savage, "Experimental Security Analysis of a Modern Automobile", IEEE Symposium on Security and Privacy, Oakland, CA, May 16–19, 2010.

[5] OVERSEE, "Open Vehicular Secure Platform", https://www.oversee-project.com.

Automotive and Industrial Data Security

André Weimerskirch

Cybersecurity for Cyber-Physical Systems Workshop

April 23-24, 2012

escrypt Inc. – Embedded Security
315 E Eisenhower Parkway, Suite 214
Ann Arbor, MI 48108, USA
+1-734-418-2797 info@escrypt.com

Overview

- **Introduction and Motivation**
- Risk analysis
- Current and future security solutions
- Conclusions

04/23/2012   Automotive and Industrial Data Security   escrypt Inc. – Embedded Security
315 E Eisenhower Parkway, Suite 214
Ann Arbor, MI 48108, USA
+1-734-418-2797 info@escrypt.com

Communication and Cars

- "If vehicles had developed in the same manner as telecommunications, then an average car would reach top speeds of 10^9 km/h at 400 million horse power, and the car would be hacked four times per year."
  - Prof. Christof Paar

04/23/2012   Automotive and Industrial Data Security   escrypt Inc. – Embedded Security
315 E Eisenhower Parkway, Suite 214
Ann Arbor, MI 48108, USA
+1-734-418-2797 info@escrypt.com

Digital revolution in vehicles

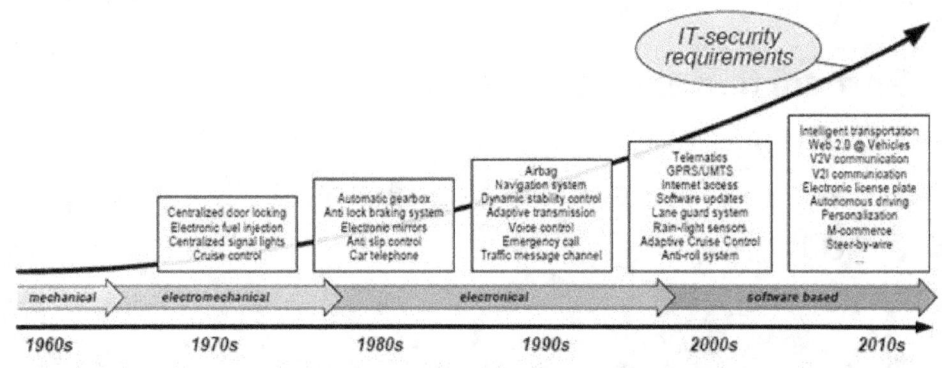

- Vehicles changed from mechanical to software based systems
- Software and electronics accounts for up to 50% of total cost
- Software and electronics is a market distinction today
- Modern cars come with up to 80 CPUs, 2 miles of cable, several hundred MB of software, and 5 in-vehicle networks

04/23/2012   Automotive and Industrial Data Security   escrypt Inc. – Embedded Security
315 E Eisenhower Parkway, Suite 214
Ann Arbor, MI 48108, USA
+1-734-418-2797 info@escrypt.com

Hacking

- Recent media reports suggest that data security in vehicle becomes an issue
  - Remote attack to vehicle by manipulated MP3 file via Bluetooth connected cell phone
  - Weakness in telematics module to gain remote access via cellular connection
  - Weakness in Bluetooth stack to gain access to CAN
  - Then remotely flash new firmware, e.g. for brake ECU: lock left rear brake once car reaches 70 mph. Attacker can be on another continent.

Trojan-Horse MP3s Could Let Hackers Break Into Your Car Remotely, Researchers Find

Source: http://www.popsci.com/cars/article/2011-03/bluetooth-music-and-cell-phones-could-let-hackers-break-your-car-researchers-say

escrypt Inc. -- Embedded Security
315 E Eisenhower Parkway, Suite 214
Ann Arbor, MI 48108, USA
+1-734-418-2797 info@escrypt.com

Hacking

- Same researchers found local attack
  - Connect laptop to OBD-II port and flash manipulated firmware
  - Insert manipulated CD or USB flash drive to inject manipulated firmware
- Remotely mounted hacks via Internet to remote engine start and remote vehicle unlock have been demonstrated as well by other researchers.
- Knowledge is proprietary and no known attacks are known. However, it is a matter of time until the knowledge will leak.

escrypt Inc. -- Embedded Security
315 E Eisenhower Parkway, Suite 214
Ann Arbor, MI 48108, USA
+1-734-418-2797 info@escrypt.com

Financial Damage

- Counterfeit black market is a gigantic problem

- Odometer rollback
  - 6 billion Euro damage per year in Germany
  - 10-30% of all sold used vehicles manipulated in USA

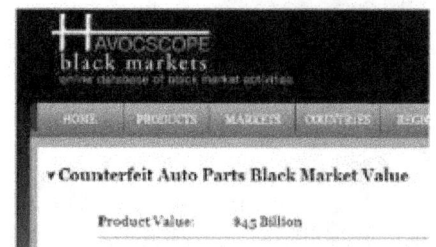

Source: http://www.havocscope.com/black-market/counterfeit-goods/counterfeit-auto-parts/

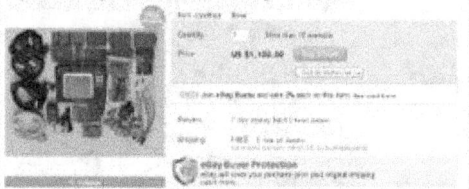

Source: http://www.ebay.com

Financial Damage

- Warranty fraud
  - Owner performs chip tuning to increase engine power
  - Engine blasts and owner flashes original firmware

Source: http://www.ebay.com

Privacy

escry̍pt
Embedded Security

- Event Data Recorder (EDR) record information during crashes and accidents.
- Navigation units include list of recent targets
- Theft protection devices track vehicles via GPS

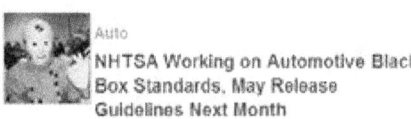

Auto

NHTSA Working on Automotive Black Box Standards, May Release Guidelines Next Month

User privacy and costs regarding the integration of high-tech EDRs are the largest concerns

The National Highway Traffic Safety Administration may make event data recorders, or "black boxes," a requirement for all vehicles starting next month, according to mired's AutoBU.

Event data recorders (EDR) are devices already installed in some automobiles, and record information during vehicle crashes or accidents. EDRs cannot be turned off, and once electronically triggered by problems in the engine or dramatic shifts in wheel speed, the EDR records that vehicle input and produces a snapshot of the final moments before the accident.

Source:
http://www.dailytech.com/NHTSA+Working+on+Automotive+Black+Box+Standards+May+Release+Guidelines+Next+Month/article21717.htm

04/23/2012   Automotive and Industrial Data Security

escrypt Inc. – Embedded Security
315 E Eisenhower Parkway, Suite 214
Ann Arbor, MI 48108, USA
+1-734-418-2797 info@escrypt.com

Area full of Pitfalls: Aftermarket

escry̍pt
Embedded Security

Hacker Disables More Than 100 Cars Remotely

By Kevin Poulsen | March 17, 2010 | 1:02 pm | Categories: Breaches, Crime, Cybersecurity, Hacks and Cracks

More than 100 drivers in Austin, Texas found their cars disabled or the horns honking out of control, after an intruder ran amok in a web-based vehicle-immobilization system normally used to get the attention of consumers delinquent in their auto payments.

Police with Austin's High Tech Crime Unit on

Source: http://www.wired.com/threatlevel/2010/03/hacker-bricks-cars/

- Be careful who "upgrades" your car!

Dutch Police Used TomTom's GPS Data To Target Speeders

Categories: Technology, Foreign News

By EYDER PERALTA

Source: http://www.npr.org/blogs/thetwo-way/2011/04/28/135809709/dutch-police-used-tomtoms-gps-data-to-target-speeders?sc=17&f=1019

04/23/2012   Automotive and Industrial Data Security

escrypt Inc. – Embedded Security
315 E Eisenhower Parkway, Suite 214
Ann Arbor, MI 48108, USA
+1-734-418-2797 info@escrypt.com

Underlying Problem: Vehicle Architecture

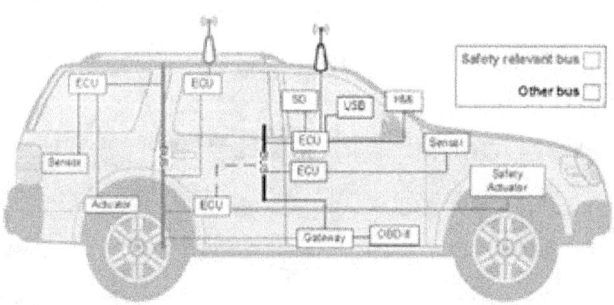

- There is a physical connection between safety relevant bus and non-safety bus
  - E.g. speed adjusted volume
  - Note the physical connection between cellular, USB, SD, HMI, and safety critical powertrain components
- The complexity of code increases, mainly due to infotainment
  - Almost impossible to avoid security flaws
- Access to bus via OBD-II, or remotely via infotainment system

04/23/2012   Automotive and Industrial Data Security   escrypt Inc. – Embedded Security
315 E Eisenhower Parkway, Suite 214
Ann Arbor, MI 48108, USA
+1-734-418-2797 info@escrypt.com

Why is data security in vehicles special?

- Safety critical: a hacked vehicle might be different than a hacked PC
- Vehicles cannot regularly update software
- In many instances, attacker has physical access

- ➤ More infotainment will be introduced
- ➤ Modern cars include 100 million lines of code
  - Industry average is about one security flaw per 1,000 lines of code
  - ➤ Around 100,000 flaws?

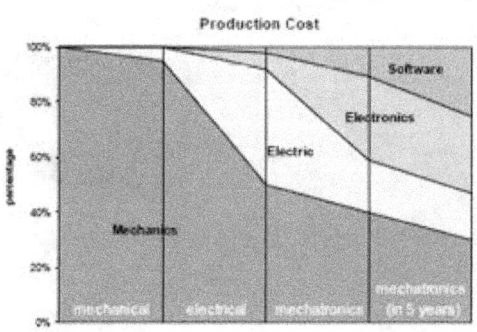

Source: Audi

04/23/2012   Automotive and Industrial Data Security   escrypt Inc. – Embedded Security
315 E Eisenhower Parkway, Suite 214
Ann Arbor, MI 48108, USA
+1-734-418-2797 info@escrypt.com

Why is automotive security different to PC security (and hard)?

- Combination of safety and infotainment
  - Buyers demand modern connected infotainment systems
  - If a PC is hacked, data is lost. If a car is hacked, life is at stake.
- Automotive software cannot easily be updated
  - No monthly security update
- Attacker might have physical access to vehicle

escrypt Inc. – Embedded Security
315 E Eisenhower Parkway, Suite 214
Ann Arbor, MI 48108, USA
+1-734-418-2797 info@escrypt.com

Safety and Security

Safe and reliable Operation

well understood – processes are in place

relatively new challenge since the introduction of wireless communication and advanced infotainment

Reliability

Protection against defects

Security

Protection against targeted hackers

escrypt Inc. – Embedded Security
315 E Eisenhower Parkway, Suite 214
Ann Arbor, MI 48108, USA
+1-734-418-2797 info@escrypt.com

Overview

- Introduction and Motivation
- **Risk analysis**
- Current and future security solutions
- Conclusions

04/24/2012 Automotive and Industrial Data Security

escrypt Inc. – Embedded Security
315 E Eisenhower Parkway, Suite 214
Ann Arbor, MI 48108, USA
+1-734-418-2797 info@escrypt.com

Who are the attackers?

- Today: different to iPhone hacker community (challenge, curiosity), similar to Pay-TV hackers; purely financial motivation
- Almost all attackers are active in black market
  - Illegal organizations
  - Mainly financial motivation
  - Significant financial damage

- Some individual "attackers" are motivated by curiosity
  - To turn off "annoying" seat belt warning
  - Turn off TV lock
  - Academic teams
  - ➢ Any damage?

04/23/2012 Automotive and Industrial Data Security

escrypt Inc. – Embedded Security
315 E Eisenhower Parkway, Suite 214
Ann Arbor, MI 48108, USA
+1-734-418-2797 info@escrypt.com

Who are the attackers? (continued)

- The attacks are implemented and sold by black market organizations
- The *user* of the attack is in almost all cases the vehicle owner
  - Odometer rollback
  - Chip tuning
  - Buyer of cheap counterfeits (did you ever buy an original car key for $250)?

- Question:
  - Will there be attackers to mount safety-critical attacks?
  - How will the attacks be offered/distributed?

04/23/2012   Automotive and Industrial Data Security

escrypt Inc. – Embedded Security
315 E Eisenhower Parkway, Suite 214
Ann Arbor, MI 48108, USA
+1-734-418-2797 info@escrypt.com

Who is damaged?

- In many cases the buyers of used cars
  - Odometer rollback
  - Chip tuning (shortens engine life-span)
  - Counterfeits (shorter life-span)
  - Stolen vehicles (either direct damage, or damage due to increased insurance rates)
- Only a few cases where car makers are damaged directly
  - Counterfeits: lost sales
  - Potentially fines by EPA: if it becomes known that chip tuning is very easy, and that engine after chip tuning violates emission regulations
  - Warranty fraud: engine burns out after chip tuning during warranty period

04/23/2012   Automotive and Industrial Data Security

escrypt Inc. – Embedded Security
315 E Eisenhower Parkway, Suite 214
Ann Arbor, MI 48108, USA
+1-734-418-2797 info@escrypt.com

Who is damaged? (continued)

- Indirect damage for car makers might be significant
  - Lost sales if insurance rates due to high theft are significantly higher than of competitors
  - Lost sales if there is negative press

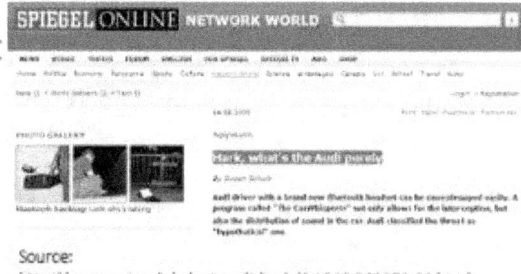

Source:
http://www.spiegel.de/netzwelt/tech/0,1518,368070,00.html

- ➢ Some pressure on car makers to introduce security

04/23/2012   Automotive and Industrial Data Security

escrypt Inc. – Embedded Security
315 E Eisenhower Parkway, Suite 214
Ann Arbor, MI 48108, USA
+1-734-418-2797 info@escrypt.com

Overview

- Introduction and Motivation
- Risk analysis
- **Current and future security solutions**
- Conclusions

04/23/2012   Automotive and Industrial Data Security

escrypt Inc. – Embedded Security
315 E Eisenhower Parkway, Suite 214
Ann Arbor, MI 48108, USA
+1-734-418-2797 info@escrypt.com

Examples of IT-Security Applications in Vehicles

- Theft protection
- Remote unlock
- Keyless entry
- Odometer manipulation
- Vehicle tracking
- Bluetooth and Wi-Fi
- eCall
- Tolling
- Business models
  - Feature activation
  - License agreements
  - Copyright protection
- Warranty: prove manipulation of firmware
- Counterfeiting of components and spare parts
- Vehicle-to-vehicle communication
- ...

escrypt Inc. – Embedded Security
315 E Eisenhower Parkway, Suite 214
Ann Arbor, MI 48108, USA
+1-734-418-2797 info@escrypt.com

Security Features: Today

- Secure flash programming
  - OEM signs firmware (usually RSA)
  - ECU verifies OEMs signature
  - Built-in public key in ECU that can be replaced with new boot-loader
  - Certificate based systems are going to be implemented

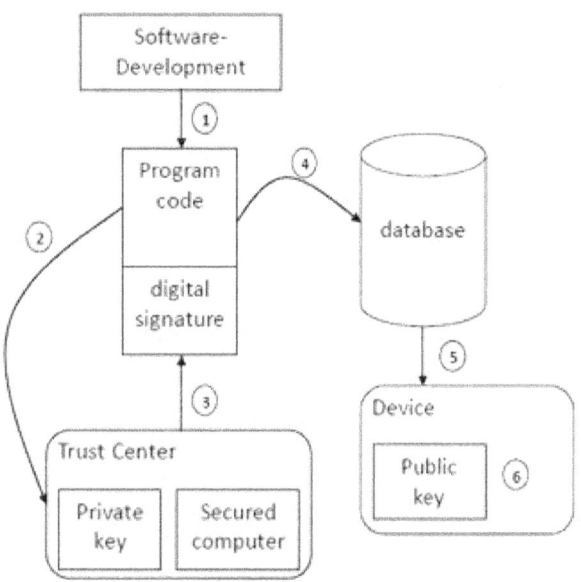

escrypt Inc. – Embedded Security
315 E Eisenhower Parkway, Suite 214
Ann Arbor, MI 48108, USA
+1-734-418-2797 info@escrypt.com

Security Features: Today

- Filter in gateway
  - Whitelists: designer explicitly define which packets are relayed between different bus systems
- Plausibility checks
  - Part of safety validation
  - Each ECU checks whether input is reasonable
  - If not, input is discarded and fail-safe mode is activated

04/23/2012   Automotive and Industrial Data Security   escrypt Inc. – Embedded Security
315 E Eisenhower Parkway, Suite 214
Ann Arbor, MI 48108, USA
+1-734-418-2797 info@escrypt.com

Security Features: Today

- Standard security
  - E.g. Bluetooth security based on PIN entered pairing
- Proprietary security
  - Theft protection
  - Feature activation

- ➢ Most OEMs focus on remote attacks

04/23/2012   Automotive and Industrial Data Security   escrypt Inc. – Embedded Security
315 E Eisenhower Parkway, Suite 214
Ann Arbor, MI 48108, USA
+1-734-418-2797 info@escrypt.com

Future In-Vehicle Security Layers

escrypt
Embedded Security

- Need for layered in-vehicle security:

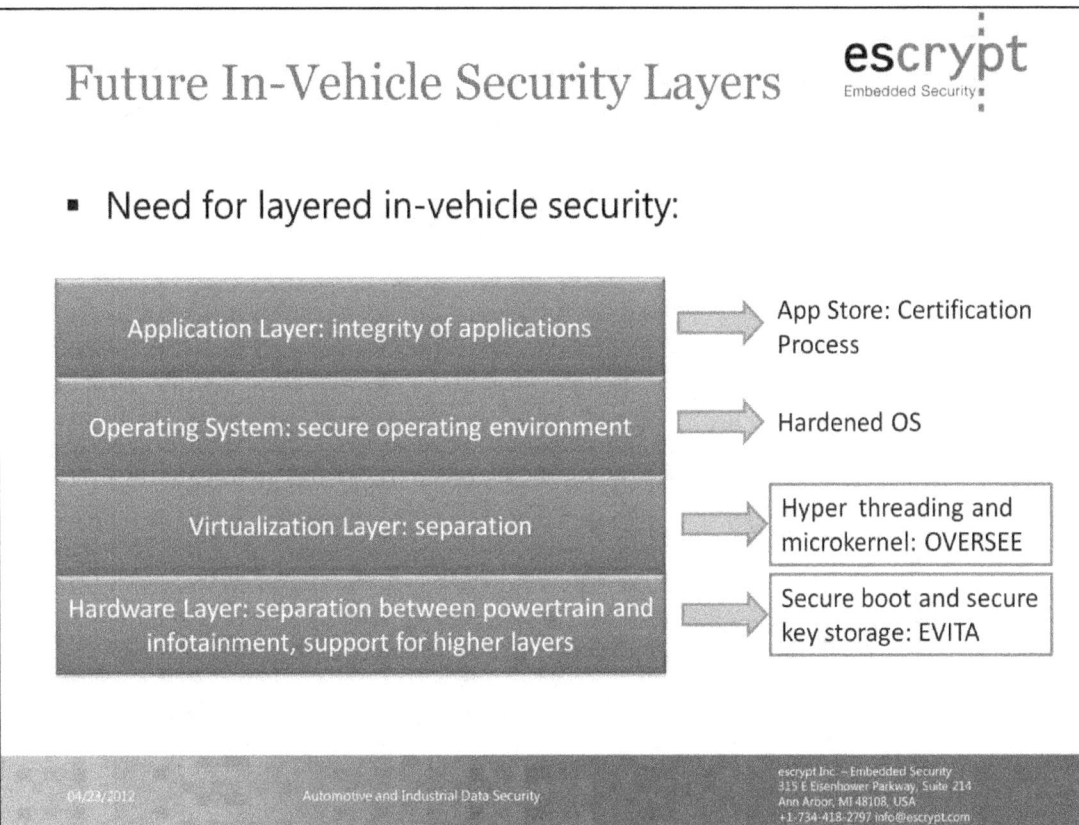

04/23/2012   Automotive and Industrial Data Security

escrypt Inc. – Embedded Security
315 E Eisenhower Parkway, Suite 214
Ann Arbor, MI 48108, USA
+1-734-418-2797 info@escrypt.com

Virtualization and Microkernel

escrypt
Embedded Security

- Rule of thumb: one security weakness per 1,000 lines of code
- Linux and Windows have many million lines of code
  - Thousands of security weaknesses
- Design microkernel
  - Remove drivers and non-essential modules from kernel
  - Between 10,000 and 100,000 lines of code
  - Hope: find all security weaknesses before deployment
  - Hope: formal verification of correctness

04/23/2012   Automotive and Industrial Data Security

escrypt Inc. – Embedded Security
315 E Eisenhower Parkway, Suite 214
Ann Arbor, MI 48108, USA
+1-734-418-2797 info@escrypt.com

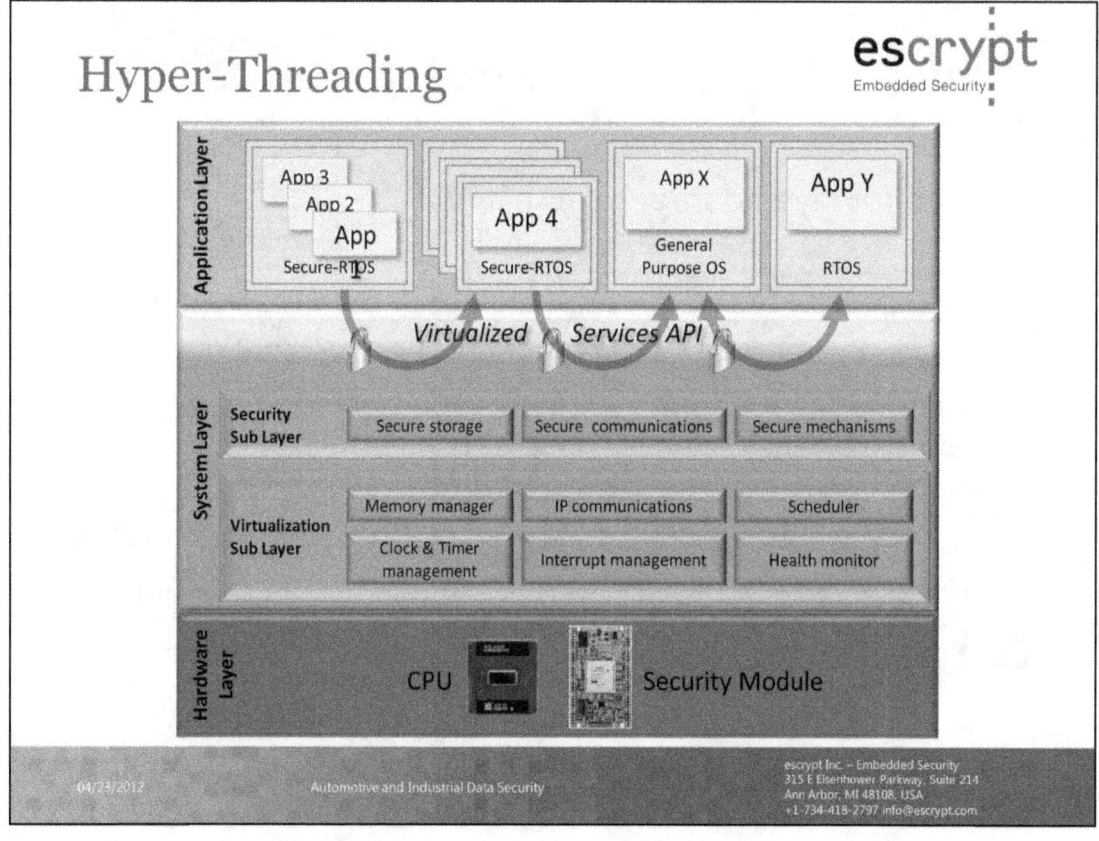

Approach: OVERSEE

- OVERSEE: Open Vehicular Secure Platform
- Objective: Providing a standardized generic communication and application platform for vehicles, ensuring security, reliability and trust of external communication and simultaneous running applications.
- European Union funded project (3 million EURO)
- Runtime 2010 – 2012
- Members: Volkswagen, ESCRYPT, Fraunhofer, Trialog, Technical University Berlin, University of Valencia, Open Tech
- More information at www.oversee-project.com

Hardware Security

1. Hardware as ultimate separation between local and remote interfaces and powertrain
   - "Automotive Firewall"

2. Hardware as security anchor for higher layers
   - Protection of software manipulation
   - Secure boot
   - Secure key storage
- Fast crypto performance

EVITA Security Levels

- **EVITA Full**: V2X (one per car)

- **EVITA Medium**: for advanced ECUs (gateway, headunit, engine control)

- **EVITA Light**: for sensors, actuators, …

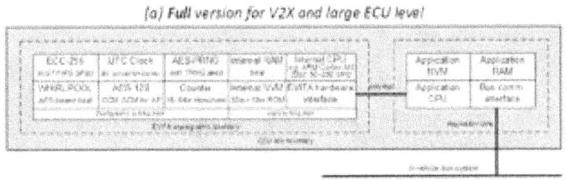

(a) Full version for V2X and large ECU level

(b) Medium version for standard ECU level

(c) Light version for sensor/actuator ECU level

EVITA in Cars

- EVITA provides hardware extensions
    - Just some more chip area
    - Cost very low
- EVITA does not provide security known from dedicated security controllers such as smart-cards
    - Only basic tamper resistance

- EVITA Light was standardized as Secure Hardware Extension (SHE)
    - Available by several semiconductors
    - Automotive grade
- A controller that implements EVITA Medium was recently introduced

04/23/2012   Automotive and Industrial Data Security

escrypt Inc. - Embedded Security
315 E Eisenhower Parkway, Suite 214
Ann Arbor, MI 48108, USA
+1-734-418-2797 info@escrypt.com

Security of Future Vehicle Application: V2X

- Vehicles are equipped with Wi-Fi (but 5.9 GHz) and regularly broadcast location, speed, vehicle category, time, ...
- Receiver application creates map of environment (no line of sight necessary)
- Receiver safety application notifies driver
    - E.g. immanent crash warning
- Probably the "hottest" security application in vehicles today
    - 260 million nodes
    - Full of privacy pit holes
    - Security and safety intermix
- Security was recognized as major component and introduced from the beginning!

04/23/2012   Automotive and Industrial Data Security

escrypt Inc. - Embedded Security
315 E Eisenhower Parkway, Suite 214
Ann Arbor, MI 48108, USA
+1-734-418-2797 info@escrypt.com

Overview

- Introduction and Motivation
- Risk analysis
- Current and future security solutions
- **Conclusions**

Conclusions 1/2

- Passenger vehicles are more and more connected
  - Users demand for infotainment known from mobile phones
- Main concern is introduced by connectivity
  - Infotainment, wireless connectivity, telematics, V2X
- Recent academic attacks suggest that modern vehicles are vulnerable to serious hacker attacks
  - No actual attacks known though
  - Knowledge very proprietary
- Also a problem (but not considered here): privacy
  - E.g. tracking based on RFID air pressure sensor

Conclusions 2/2

- Car manufacturers work on solutions
  - Plenty of data security mechanisms already implemented today
  - Powertrain needs to be efficiently separated from external communication channels
- Lots of momentum
  - Semiconductors introduce automotive security controllers
    - Secure Hardware Extension (SHE)
    - EVITA Medium (HSM)
  - US DOT discusses introduction of automotive Information Sharing and Analysis Center (ISAC)
  - SAE set up Vehicle Electrical System Security Committee

04/23/2012   Automotive and Industrial Data Security

escrypt Inc. – Embedded Security
315 E Eisenhower Parkway, Suite 214
Ann Arbor, MI 48108, USA
+1-734-418-2797 info@escrypt.com

5. Keynote Speaker: Dr. Farnham Jahanian

Dr. Farnham Jahanian
Assistant Director for CISE
National Science Foundation

Farnam Jahanian serves as the National Science Foundation *Assistant Director for the Computer and Information Science and Engineering (CISE) Directorate. He* guides CISE, with a budget of over $650 million, in its mission to uphold the nation's leadership in computer and information science and engineering through support of fundamental and transformative advances that are a key driver of economic competitiveness and that are crucial to achieving national priorities. *Dr. Jahanian is also* co-chair of the Networking and Information Technology Research and Development (NITRD) Subcommittee of the National Science and Technology Council Committee on Technology, providing overall coordination for the activities of 15 government agencies.

Dr. Jahanian is on leave from the University of Michigan, where he holds the Edward S. Davidson Collegiate Professorship and served as Chair for Computer Science and Engineering from 2007 – 2011 and as Director of the Software Systems Laboratory from 1997 – 2000. His research on Internet infrastructure security formed the basis for the Internet security company Arbor Networks, which he co-founded in 2001. He served as Chairman of Arbor Networks until its acquisition by Tektronix Communication in 2010. Dr. Jahanian holds a master's degree and a Ph.D. in Computer Science from the University of Texas at Austin. He is a Fellow of the Association for Computing Machinery (ACM), the Institute of Electrical and Electronic Engineers (IEEE), and the American Association for the Advancement of Science (AAAS).

Secure Smart Systems:
A Global Imperative

Farnam Jahanian
CISE Directorate
National Science Foundation

NIST Cybersecurity for Cyber-Physical Systems Workshop
April 23, 2012

Smart Infrastructure

Imagine a day where...
static infrastructure is adaptable and safe

Image Credit: MicroStrain, Inc.

Environment and Sustainability

Imagine a day where...
   we can forecast and mitigate ecological change

Health and Wellbeing

Imagine a day where...
   wellbeing is pervasive and healthcare is personalized

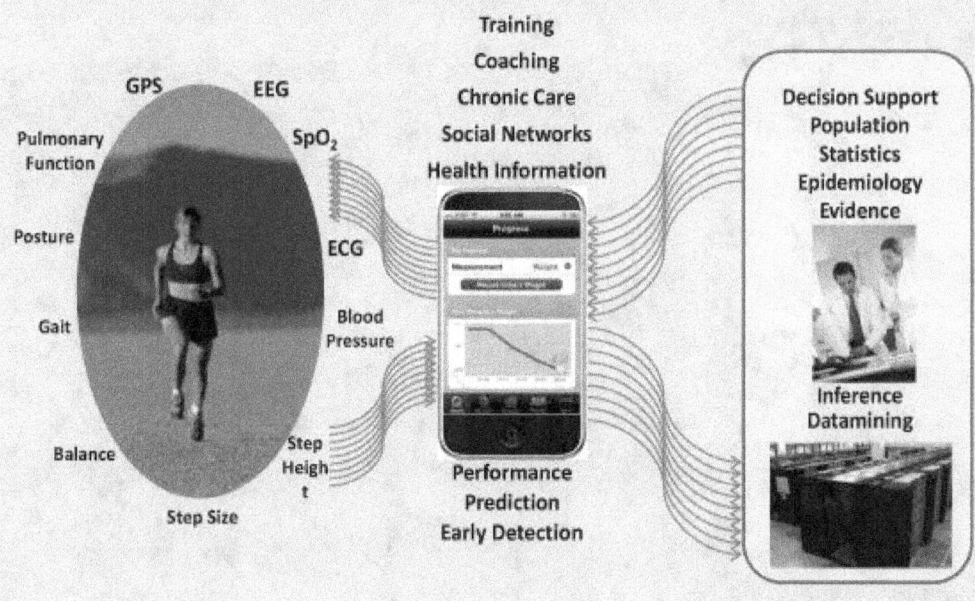

Smart Grids

Imagine a day where...
energy is efficiently used and intelligently managed

Emergency Response

Imagine a day where...
we can prevent, mitigate, and recover from disasters

Image Credits: Karen Gessy, NSF (left) and Texas A&M University (right)

Transportation: Safety and Energy

Imagine a day where...
traffic fatalities no longer exist

The Promise

Advances in *cyber-physical systems* hold the potential to reshape our world with more responsive, secure, and efficient systems that:

- transform the way we live
- drive economic prosperity
- underpin national security
- enhance societal well-being

CPS and National Priorities

| Manufacturing, Robotics, & Smart Systems | Environment & Sustainability | Emergency Response & Disaster Resiliency | Health & Wellbeing |

| Transportation & Energy | Broadband & Universal Connectivity | Secure Cyberspace | Education and Workforce Development |

A National Imperative

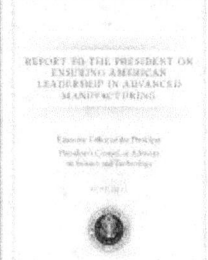

- **2007 PCAST NITRD Report** – Recommended cross-disciplinary programs to accelerate work in CPS by Federal R&D agencies

- **2010 PCAST NITRD Report** – Expanded this recommendation to energy, transportation, health care, and homeland security

- **2011 PCAST Advanced Manufacturing Report** – Recommended investments to strength US leadership in the areas of robotics, cyber-physical systems, and flexible manufacturing

Cyber-Physical Systems

Deeply integrating computation, communication, and control into physical systems

- Pervasive computation, sensing and control
- Networked at multi- and extreme scales
- Dynamically reorganizing/reconfiguring
- High degrees of automation
- Dependable operation with high assurance of reliability, safety, security and usability

Transportation
- Faster and safer aircraft
- Improved use of airspace
- Safer, more efficient care

Energy and Industrial Automation
- Homes and offices that are more energy efficient and cheaper to operate
- Distributed micro-generation for the grid

Healthcare and Biomedical
- Increased use of effective in-home care
- More capable devices for diagnosis
- New internal and external prosthetics

Critical Infrastructure
- More reliable power grid
- Highways that allow denser traffic with increased safety

Smart Systems: Sensing, Reasoning, and Decision

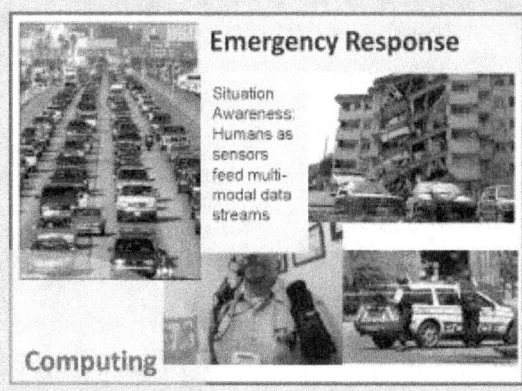

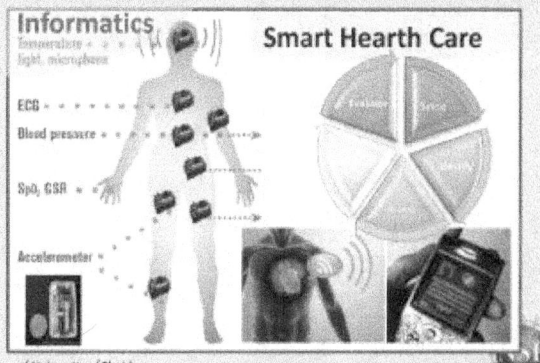

Source: Sajal Das, Keith Marzullo   Image Credit: Image courtesy of University of Florida

A Sea of Sensors

- We swim in **a sea of sensors and are drowning in data**:
  - Ability to analyze data in **real-time** and **retrospectively;** to create context for decisions; and to offer meaningful actionable feedback
  - As called for in the 2010 PCAST report, networked systems that not only **scale up**, but also **scale down** and **scale out**:
    - Smart, miniaturized, low-power, adaptive and self-calibrating instrumentation
    - Embedded sensors everywhere and connecting everything via networks leading to wide-scale sensing and control

- Research challenges:
  - Develop new scientific and engineering principles, algorithms, models, and theories for the analysis and design of CPS.
  - How do we build systems that combine the cyber and the physical world? Abstract representation of the physical world? Models for interaction w/ the physical world?

Realizing the Potential of CPS

- Establish a scientific basis for CPS: unified foundations, models, tools, and principles

- Synthesize knowledge from disciplines that interface the cyber and physical worlds to model and simulate complex systems and dynamics

- Enable usability, adoption, and deployment of complex systems through fundamental cognitive, behavioral, economic, social, and decision sciences

- Design for reliable, robust, safe, scalable, secure, and certifiably dependable control of complex systems – CPS people can bet their lives on
  - Support networked, cyber-physical systems with built-in assurance, safety, security, and predictable performance

- Develop, document, and disseminate research-based standards and best practices for CPS

- Advance cyber-enabled discovery and innovation to enhance understanding and management of complex systems

- Prepare the next generation of talent for CPS through education and workforce development

Enable a research community and workforce that will be prepared to address the challenges of next generation systems	Bridge previously separated areas of research to develop a unified systems science for cyber-physical systems	Develop new educational strategies for a 21st century CPS workforce that is conversant in both cyber and physical aspects of systems

NSF CPS Awards Span Many Sectors

Assistive Medical Technologies: Programmable second skin senses and re-educates injured nervous systems. (Eugene Goldfield, Harvard Medical School)

 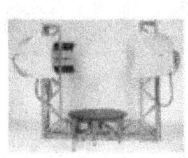 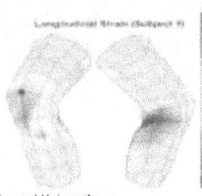 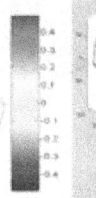 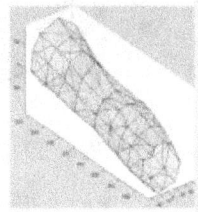

Image Credit: Wyss Institute, Harvard University

Environmental Sensing: Modeling and software allow actuated sensing in dynamic environments, such as rivers. (Jonathan Sprinkle, U. Arizona; Sonia Martinez, UCSD; Alex Bayen, UC Berkeley)

Autonomous Vehicles: Development of precision and real-time sensors, smart algorithms, and verification tools enables self-driving cars. (Ragunathan "Raj" Rajkumar, CMU, et al.)

A World of Cyber Threats

- DDoS attacks
- Worms
- Trojan Horses
- Spyware

- Botnets
- Phishing
- Insider misuse
- Data theft

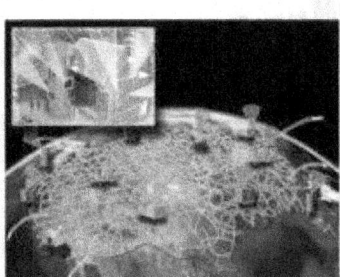

Image Credit: Nicolle Rager Fuller, NSF

How can we design, build and verify reliable, predictable, safe and **secure** cyber-physical systems upon which people can - and will - bet their lives?

Why is the Cyber Security Challenge so Difficult?

- **Attacks and defenses co-evolve**: a system that was secure yesterday might no longer be secure tomorrow.

- The technology base of our systems is frequently updated to improve functionality, availability, and/or performance. **New systems introduce new vulnerabilities** that need new defenses.

- The **environments** in which our computing systems are deployed and the functionality they provide are **dynamic**, e.g. cloud computing, mobile platforms.

- The **sophistication** of attackers is increasing as well as their sheer **number** and the **specificity** of their targets.

- As **automation pervades new platforms**, vulnerabilities will be found in critical infrastructure, automotive systems, medical devices.

- Cyber security is a **multi-dimensional** problem requiring expertise from CS, mathematics, economics, behavioral and social sciences.

The Early Years: *'Cyber Vandalism'*

- Primary motivation of hackers was bragging rights

- Worms and viruses intended to simply wreak havoc on the infrastructure

- These were availability attacks: impacting network access and services, and often, reputations

The Rise of Botnets: Cyber Crime

- Dramatic Transformation and Escalation
  - A compromised system is more useful alive than dead
  - A compromised system provides anonymity
  - A network of compromised hosts provides a powerful delivery platform
- Botnets represent today's attack platform
- Botnets will continue to dominate how attacks are launched

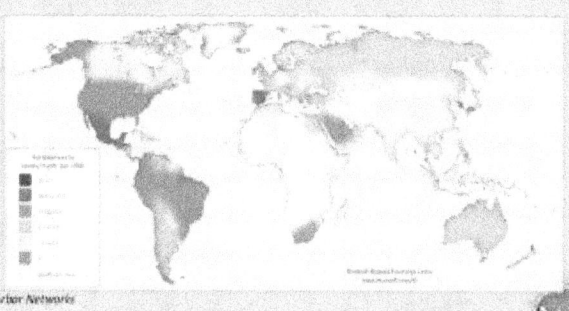

Rank	Family	Primary Control Mechanism	Computers Cleaned (1Q10)	Computers Cleaned (2Q10)	Change
1	Win32/Rimecud	Other	1,807,773	1,748,260	-3.3% ▼
2	Win32/Alureon	HTTP	1,463,885	1,035,079	-29.3% ▼
3	Win32/Hamweq	IRC	1,117,380	779,731	-30.2% ▼
4	Win32/Pushbot	IRC	474,761	589,248	24.1% ▲
5	Win32/IRCbot	IRC	597,654	388,749	-35.0% ▼

Image Credit: Arbor Networks

Microsoft desktop anti-malware products removed bots from 6.5 million computers around the world in 2Q10 [Microsoft SIR v9]

Cheaper, Better, Faster

- While Internet threat complexity is increasing at a dramatic rate, costs are dropping

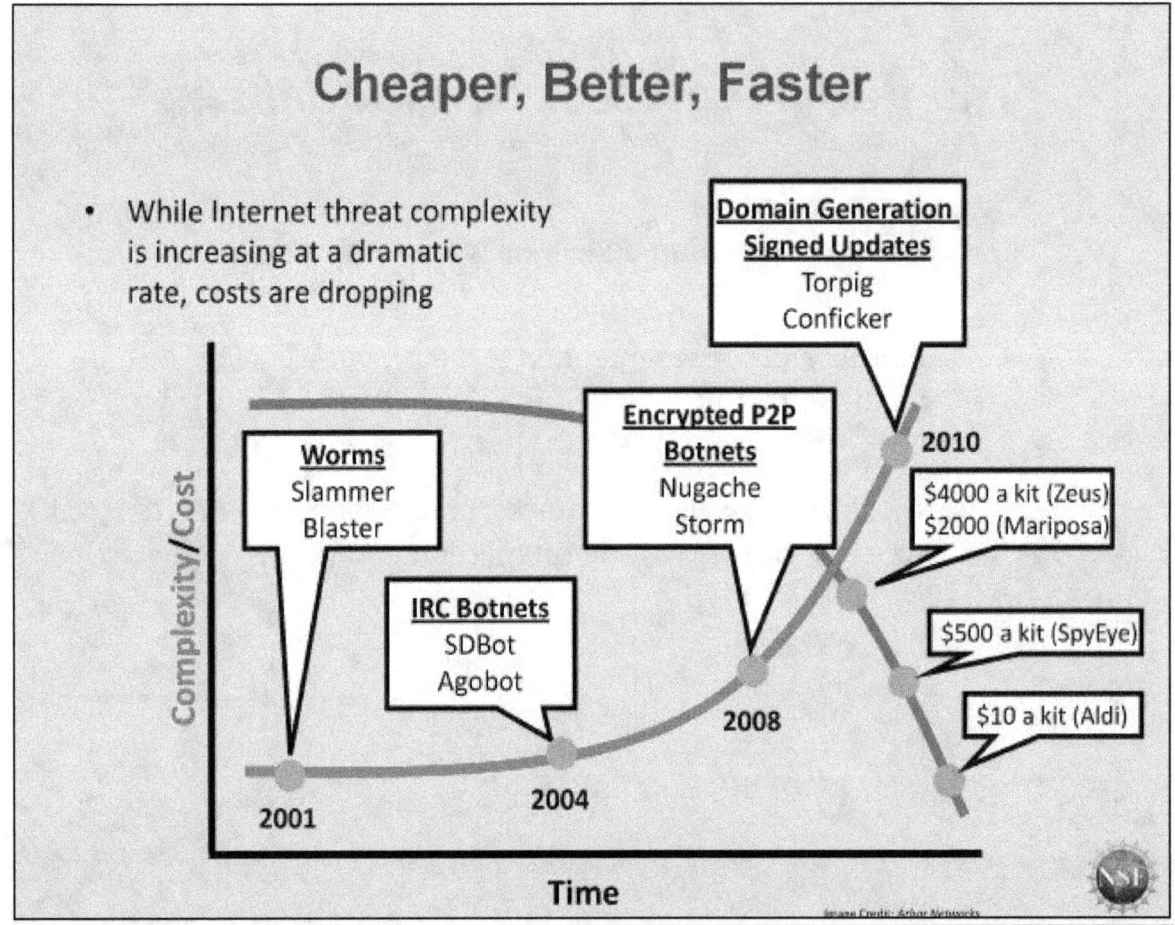

Increasing Size, Sophistication, Targeting

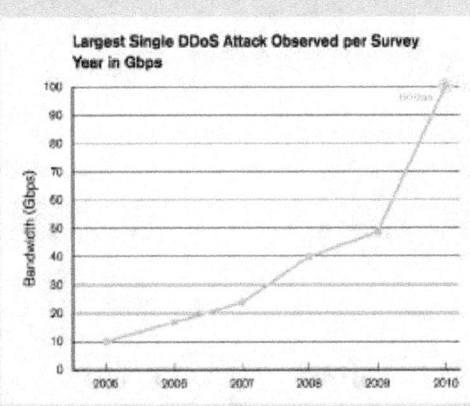

- Increasing size and sophistication
- 10-100Gbps DDoS attacks seen by ISPs
- Attacks moving "up" – Attacking services rather than infrastructure

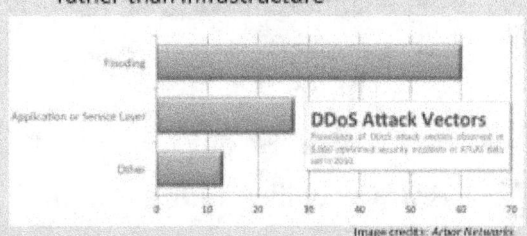

- Exploits moving "up" as well – infections now delivered via web sites through drive-by installs
  - Projected 1 in 10 web sites hosts malicious content
  - Web-based delivery means outpacing email, viruses, etc.

Cyber-war, Censorship, Activism, and the Rise of Politically Motivated Attacks

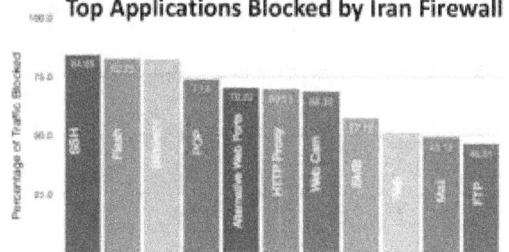

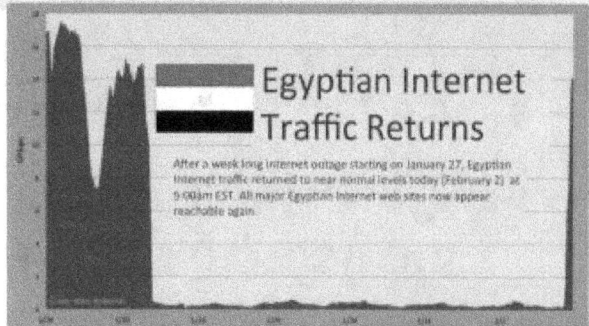

Egyptian Internet Traffic Returns

After a week long internet outage starting on January 27, Egyptian internet traffic returned to near normal levels today (February 2) at 9:00am EST. All major Egyptian internet web sites now appear reachable again.

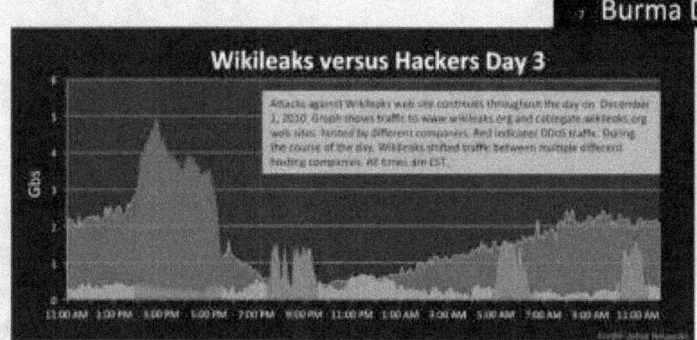

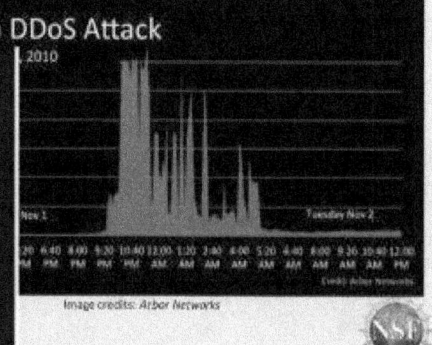

Burma DDoS Attack

Image credits: Arbor Networks

Evolution of Cyber Threats

Future security challenges will follow technology & Internet adoption patterns:

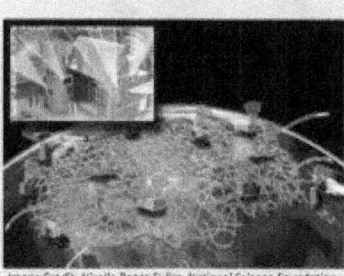

Image Credit: Nicolle Rager Fuller, National Science Foundation

- Botnets will continue to dominate how attacks are launched; attribution and forensics is increasingly difficult.
- Distributed attacks increasing in size and sophistication, targeting specific applications.
- Proliferation of attacks spurred by financial gains and now political motives.
- Proliferation of wireless devices and social media platforms open new avenues for hackers.
- Protecting cloud infrastructure key to long-term adoption.
- The trend toward increasingly cyber-enabled systems expands the scope of attacks to physical infrastructure – manufacturing, energy production, healthcare and transportation.

As the trend towards increasingly
cyber-enabled systems grows, so does
the need to secure those systems.

Cyber-Physical Security Risks

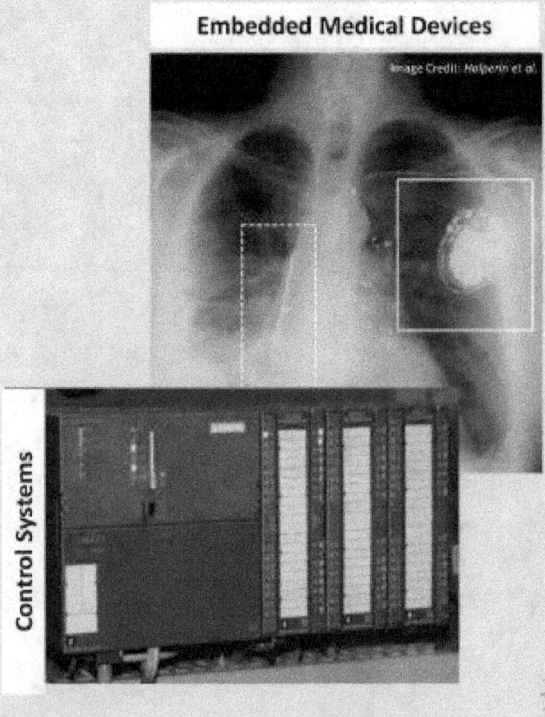

Security Risks in Automotive Computers and Networks

- Computer scientists and engineers have demonstrated ability to remotely take over automotive control systems
- In one case, by connecting to a standard diagnostic computer port included in late-model cars, caused disruption to brakes, speedometer reading, and vehicle telematics
- They are now working with the automotive industry to develop new methods for assuring the security as well as safety of automotive electronics

This car was not moving

Stefan Savage (UC San Diego) and Tadayoshi Kohno (U Washington)

Medical Device Security

As of 2006, more than half of medical devices on the US market now contain and trust software.

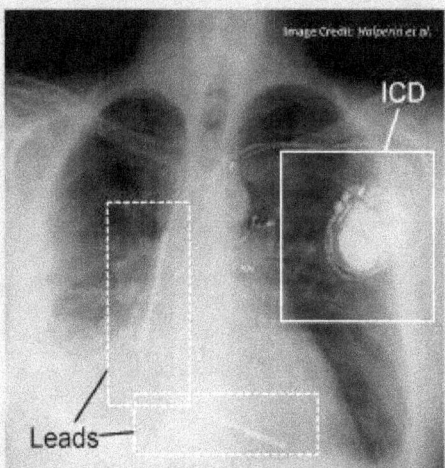

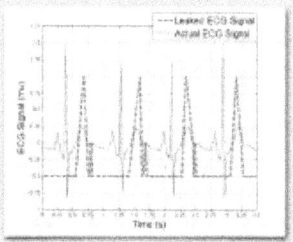

Telemedicine Privacy
[Salajegheh et al., J. Med. Dev. '09]

AED Security
[Hanna et al., HealthSec '11]

Defibrillator Vulnerabilities,
Zero-Power Defenses
[Halperin et al., IEEE S&P '08]

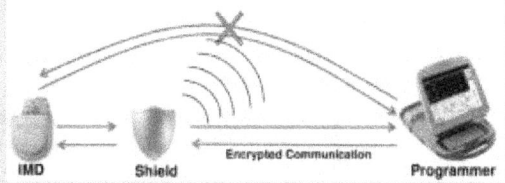

Radio Shield/Jamming for Implants
[Gollakota et al., ACM SIGCOMM '11]

Funded by NSF CNS-0435065, CNS-0520729, CNS-0627529, CNS-0831244, CNS-0842695

Implantable Medical Device Security

- Implanted medical devices frequently incorporate wireless control
- By gaining wireless access to a combination heart defibrillator and pacemaker, were able to reprogram it to shut down and to deliver jolts of electricity
- Attack vector: the device test mechanism, wireless communication interface with a control mechanism that was unencrypted
- Computer scientists working with physicians found new ways to secure these devices against extraneous signals and wireless attacks
- Encryption but also "cloakers" – make your implants "invisible" at your discretion

PI: Kevin Fu, UMass – Amherst
[Halperin et al., IEEE Symposium on Security & Privacy 2008]

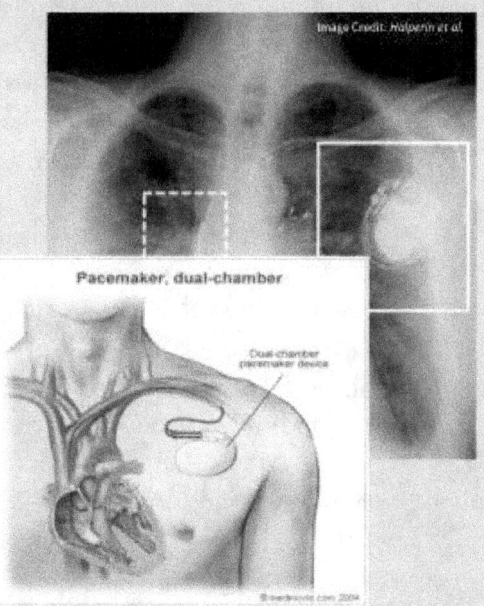

Pacemaker, dual-chamber

How Much SW in Medical Devices?

- 1983-1997
  - 6% of all recalls attributed to SW
- 1999-2005
  - **Almost doubled**: 11.3% of all recalls attributed to SW
  - 49% of all recalled devices relied on software (up from 24%)
- 1991-2000
  - **Doubled**: # of pacemakers and ICDs recalled because of SW

- 2006
  - Milestone: Over half of medical devices now involve software
- 2002-2010
  - 537+ recalls of SW-based devices affecting 1,527,311+ devices

[Sources: Biznakov et al. 2006, Fans 2006, Maisel et al. 2002, Wallace & Kuhn 2001, Computer history museum, eurogamer.net, wikipedia, businesspundit.com, imdb.com, coveitbrowser.com, thefilmtalk.com]

SCADA Security

- Targets industrial control systems, such as power plants
- Enters an organization thru an infected removable drive
- Zero-day exploits
- Anti-virus evasion techniques
- P-2-P update propagation
- Reprogramming PLC code
- Sophisticated exploitation of attack surface for a CPS

Action Webs:
Networked embedded sensor-rich systems

- Modeling, testing and validating "action webs" to achieve high-confidence networked sensor-rich control systems
- Approach: develop a theory of "action webs" using stochastic hybrid systems; taskable, multi-modal, and mobile sensor webs; and multi-scale action-perception hierarchies
  - With focus on cybersecurity: threat assessment, attack diagnosis, and resilient control
- Applications:
  - Intelligent Buildings for optimal heating, ventilation, air conditioning, and lighting based on occupant behavior and external environment
  - Air Traffic Control for mobile vehicle platforms with sensor suites for environmental sensing to enable safe, convenient, and energy efficient routing

Claire Tomlin (UC Berkeley), et al.

Design and Certification of Dependable Open Systems

- Certification and approval process are staggering
- Safety certification practice typically supports fixed configuration
- Wireless, open systems and interoperation introduce many new certification challenges

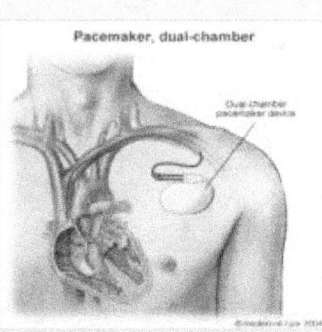

Pacemaker, dual-chamber

Improving Device Safety:
In cooperation with FDA, NSF projects are underway to design, validate, and accelerate certification of medical devices.
(University of Illinois, University of Pennsylvania, Harvard Medical/Mass General, University of Maryland, Kansas State University)

Insup Lee (U Penn), et al.

Foundations of Secure Cyber Physical Systems

- Cyber-physical systems regulating critical infrastructures, such as electrical grids and water networks, are increasingly geographically distributed, necessitating communication between remote sensors, actuators and controllers
- Combination of networked computational and physical subsystems leads to new security vulnerabilities that adversaries can exploit
- Approach: design new secure protocols and architectures for CPS through a unified conceptual framework - models for the physical system and the communication/computation network to define precise attack models and vulnerabilities
- Models are used to design protocols with provable security guarantees, thus enabling the design of more trustworthy architectures and components
- Applications: smart buildings, transportation networks, and smart grids

Image Credit: *Cisco, Inc.*

Suhas Diggavi (UCLA), et al.

Smart Grid Security

- *The CyberPhysical Challenges of Transient Stability and Security in Power Grids*
  Ian Dobson (Iowa State University), et. al.
  - Focuses on the analysis of instabilities of electric power networks, and on design of cyber-physical control methods to monitor, detect, and mitigate them
  - The controls must perform robustly in the presence of variability and uncertainty in electric generation, loads, communications, and equipment status, and during abnormal states caused by natural faults or malicious attacks

- *Trustworthy Cyber Infrastructure for the Power Grid (TCIPG) at the Univ. of Illinois*
  - Focused on securing the low-level devices, communications, and data systems that make up the power grid, to ensure trustworthy operation during normal conditions, cyber-attacks, and/or power emergencies

- *Information and Computation Hierarchy for Smart Grids* - WenZhan Song (GSU), et. al.
  - Support for high penetrations of renewable energy sources, community based micro-grids, and the widespread use of electric cars and smart appliances
  - Investigates cloud-based computing architecture for smart grids, and temporal and spatial characteristics of information hierarchy

CPS Support across NSF

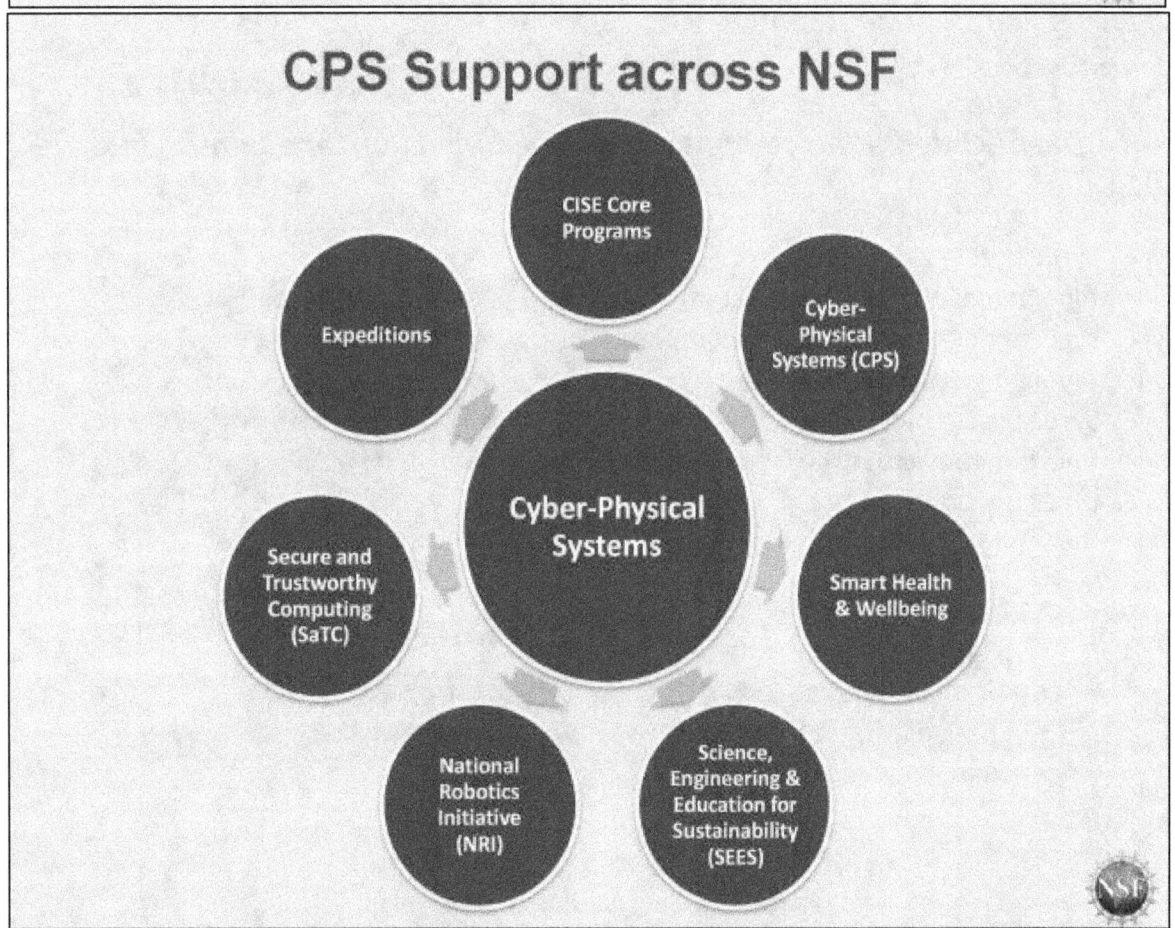

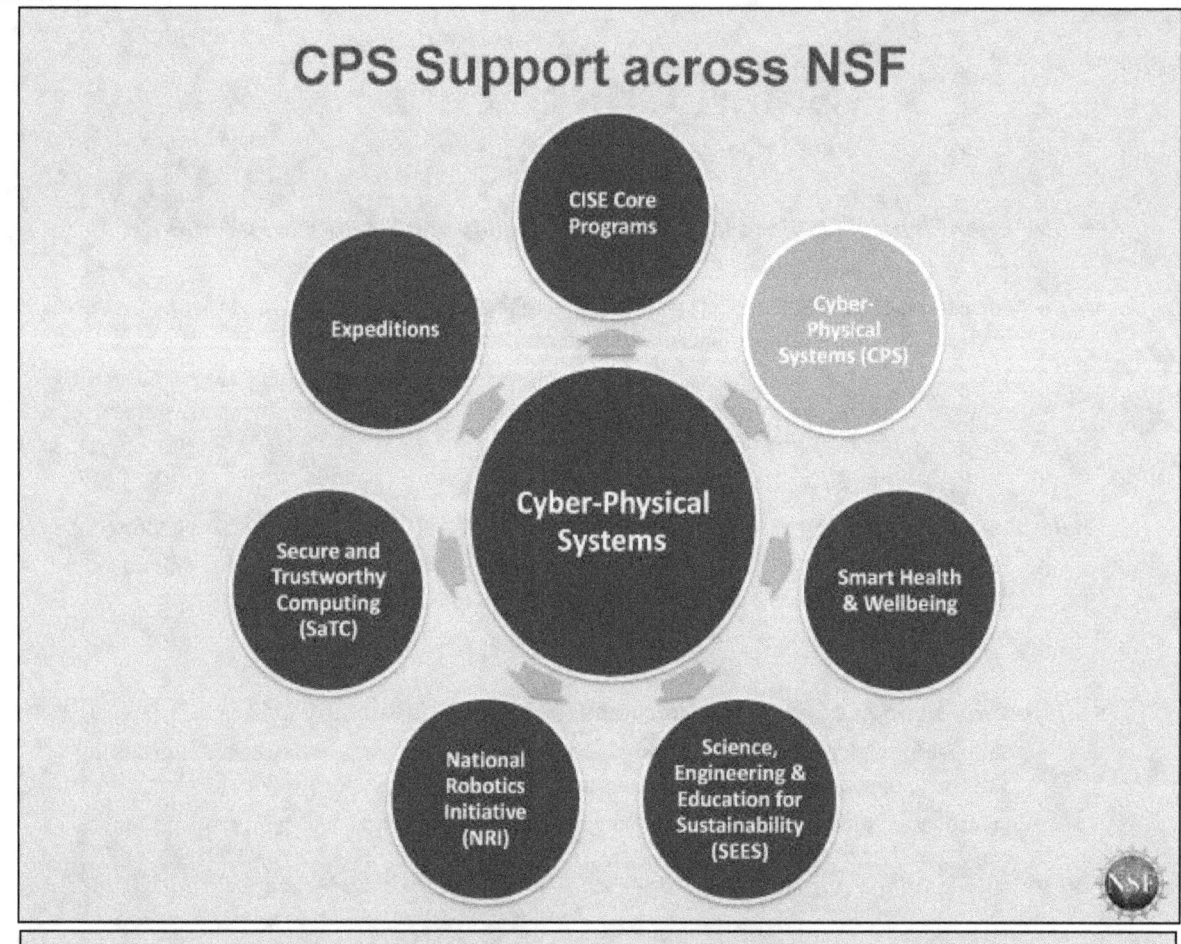

CPS Support across NSF

CISE Core Programs

Expeditions

Cyber-Physical Systems (CPS)

Secure and Trustworthy Computing (SaTC)

Cyber-Physical Systems

Smart Health & Wellbeing

National Robotics Initiative (NRI)

Science, Engineering & Education for Sustainability (SEES)

Cyber-Physical Systems Program

Deeply integrating computation, communication, and control into physical systems

- Launched in 2009
- Aims to develop the core system science needed to engineer complex "smart" cyber-physical systems
- Serves key national priorities
- Coordinated across NSF and with other government agencies

114 active awards:
- $140M+ total investment
- 43 small, average $527K
- 66 medium, average $1.5M
- 5 large, average $4.7M

Transportation

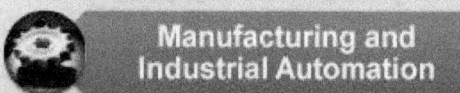
Manufacturing and Industrial Automation

Energy

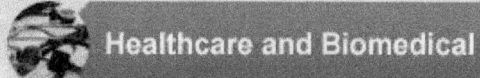
Healthcare and Biomedical

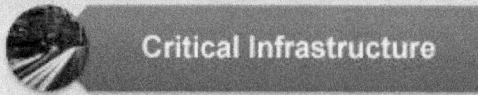
Critical Infrastructure

Cross-Directorate Solicitation: CISE and ENG

Three CPS Research Themes

- *Science* of Cyber-Physical Systems: New models and theories that unify perspectives, capable of expressing the interacting dynamics of the computational and physical components of a system in a dynamic environment. A unified science would support composition, bridge the computational versus physical notions of time and space, cope with uncertainty, and enable cyber-physical systems to interoperate and evolve.

- *Technology* for Cyber-Physical Systems: New design, analysis, and verification tools are needed that embody the scientific principles of CPS, and that incorporate measurement, dynamics, and control. New building blocks are also needed, including hardware computing platforms, operating systems, and middleware.

- *Engineering* of Cyber-Physical Systems: New opportunity to rethink principles of systems engineering, built on the foundation of CPS science and technology and able to support open cyber-physical systems. Focus on system architecture, design, integration, and design space exploration that produce certifiably dependable systems.

FY12 Solicitation

Breakthrough projects:
- Must offer a significant advance in fundamental CPS science, engineering and/or technology that has the potential to change the field
- Up to $750K for 3 yrs

Synergy projects:
- Must demonstrate innovation at the intersection of multiple disciplines, to accomplish a clear goal that requires an integrated perspective spanning the disciplines
- $750K to $2M for 3-4 yrs

Frontiers projects:
- Must address clearly identified critical CPS challenges that cannot be achieved by a set of smaller projects
- $1.2M to $10M for 4-5 yrs

CPS Virtual Organization (VO)

Objectives:
* Community building
* Technical support for collaboration
* Technology transfer and translational research
* International collaboration

http://cps-vo.org

Principles & Services:
* Community controlled
  - Information dissemination to and by the research community
* Services for collaborative activities Support for SIGs
* Industry academy interactions
* Built on open source framework
* Home for the community's historical reference materials
* Advertising of new events (e.g., calendar of upcoming events)
* Discussion forums and instant messaging
* Community members list and matchmaking

~1000 users + increasing interest by other federal agencies to provide open source results and to increase interactions among research communities

CPS Support across NSF

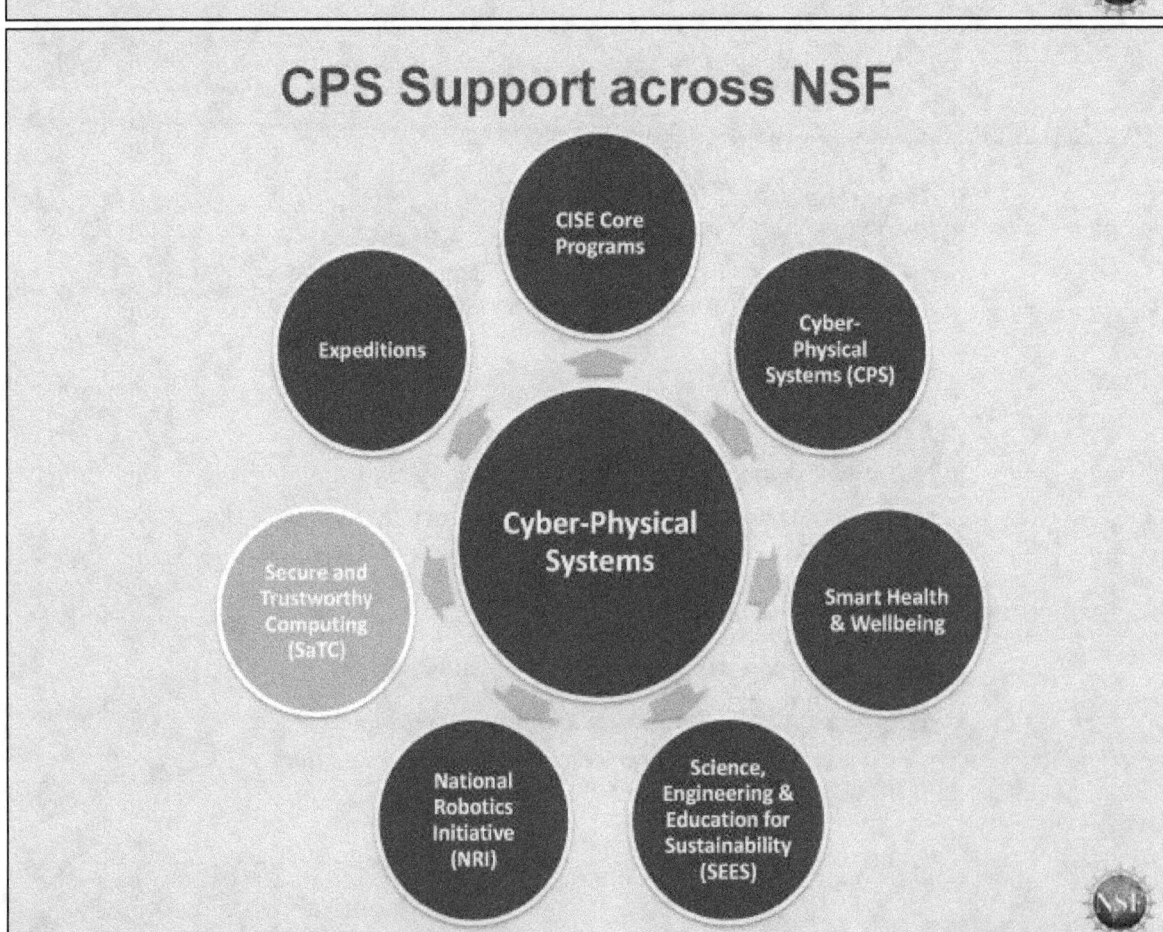

Secure and Trustworthy Cyberspace (SaTC)

Securing our Nation's cyberspace

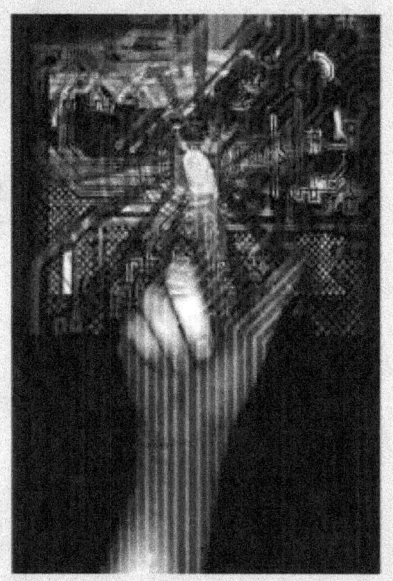

- Aims to support fundamental scientific advances and technologies to protect cyber-systems from malicious behavior, while preserving privacy and promoting usability.

- Program addresses three perspectives:
  - Trustworthy Computing Systems
  - Social, Behavioral and Economic Sciences
  - Transition to Practice

Image Credit: ThinkStock

Cross-Directorate Effort: CISE, ENG, EHR, MPS, OCI, and SBE

SaTC: Program Scope and Principles

Cast a wide net and let the best ideas surface, rather than pursuing a prescriptive research agenda

Engage the research community in developing new fundamental ideas and concepts

Promote a healthy connection between academia and a broad spectrum of public and private stakeholders to enable transition of innovative and transformative results

Project Types:

• Small	• Medium	• Frontier
up to $500,000 over 3 years	up to $1,200,000 over 4 years	up to $10,000,000 over 5 years

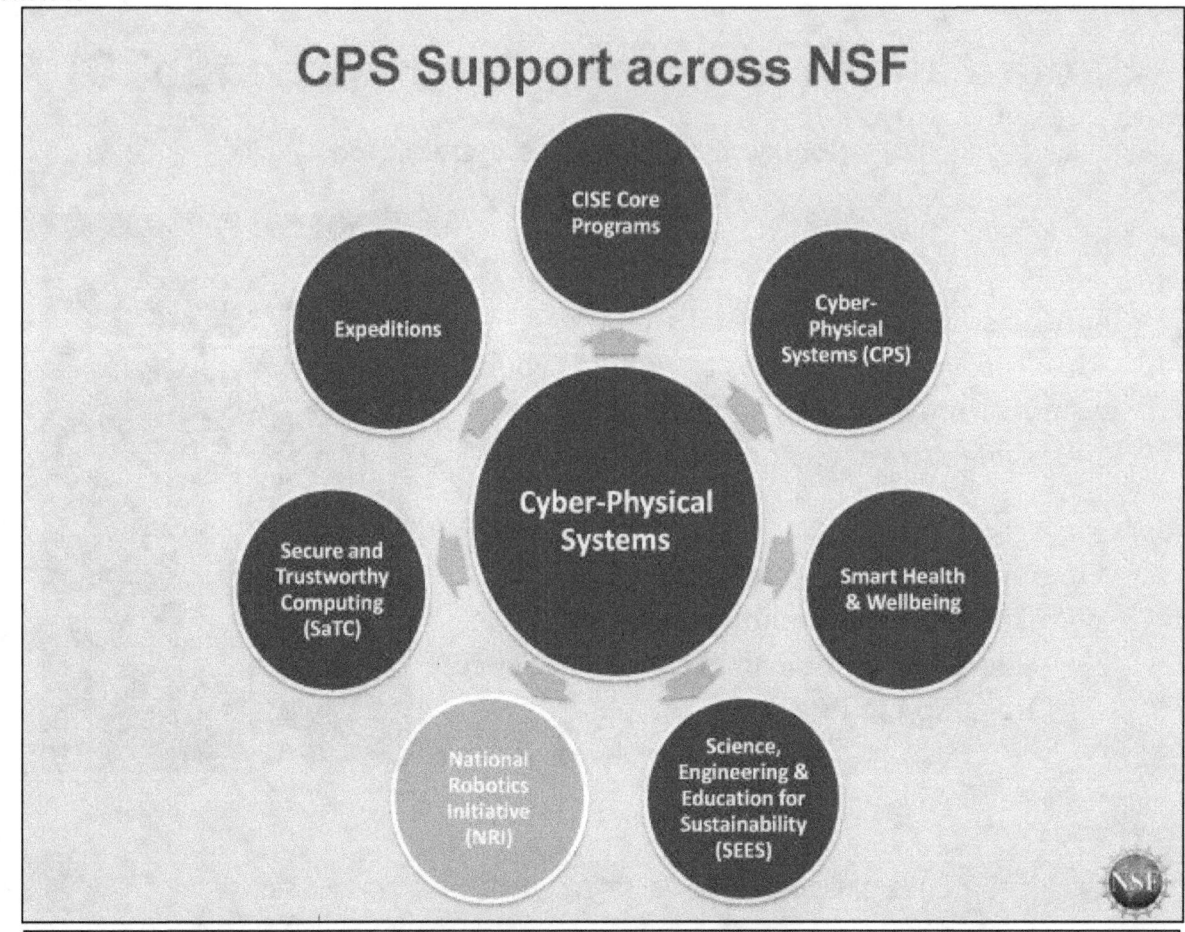

CPS Support across NSF

(CISE Core Programs; Cyber-Physical Systems (CPS); Smart Health & Wellbeing; Science, Engineering & Education for Sustainability (SEES); National Robotics Initiative (NRI); Secure and Trustworthy Computing (SaTC); Expeditions — arranged around Cyber-Physical Systems)

National Robotics Initiative (NRI)

Developing the next generation of collaborative robots to enhance personal safety, health, and productivity

A nationally concerted cross-agency program to provide U.S. leadership in science and engineering research and education aimed at the development and use of cooperative robots that work alongside people across many sectors.

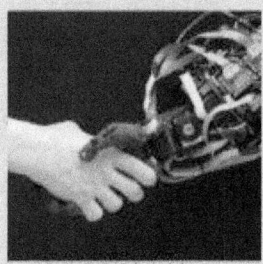

Image Credit: Bristol Robotics Lab

Research Thrusts

- Fundamental research in robotics science & engineering
- Understanding the long term social, behavioral, and economic implications across all areas of human activity
- Use of robotics to facilitate and motivate STEM learning across the K-16 continuum

Cross-Directorate Program: CISE, EHR, ENG, and SBE

Multi-agency Commitment: NSF, NASA, NIH, USDA

Wrap Up

- As automation pervades new platforms, the trend toward increasingly cyber-enabled physical infrastructure introduces new security challenges – energy production, industrial control, healthcare and transportation.

- Unsafe operation can cause significant damage to life and/or property; may pose an emerging threat to national security and defense.

- Must consider both the physical aspects of the equipment and the cyber aspects of the controls, communications, and computers that run the system:
  - Cyber-physical systems have increasing complex attack surfaces: hard to identify, measure and assess potential risk
  - Overconfidence of system designers and engineers combined with overconfidence of infrastructure operators
  - We tend to underinvest in protection and overinvest in response

Wrap Up

- We need to invest in a **research pipeline** (portfolio) comprising of long-term foundational research for secure cyber-physical systems, experimental prototypes, and early deployments to spur innovative applications.

- The CPS R&D community will continue to have a transformative and durable impact on our national priorities.

- NSF is committed to foster this emerging, consolidating research community and to reinforce its sustained role in advancing frontiers of science and engineering innovation.

- A vibrant discovery and innovation ecosystem is critical to success.

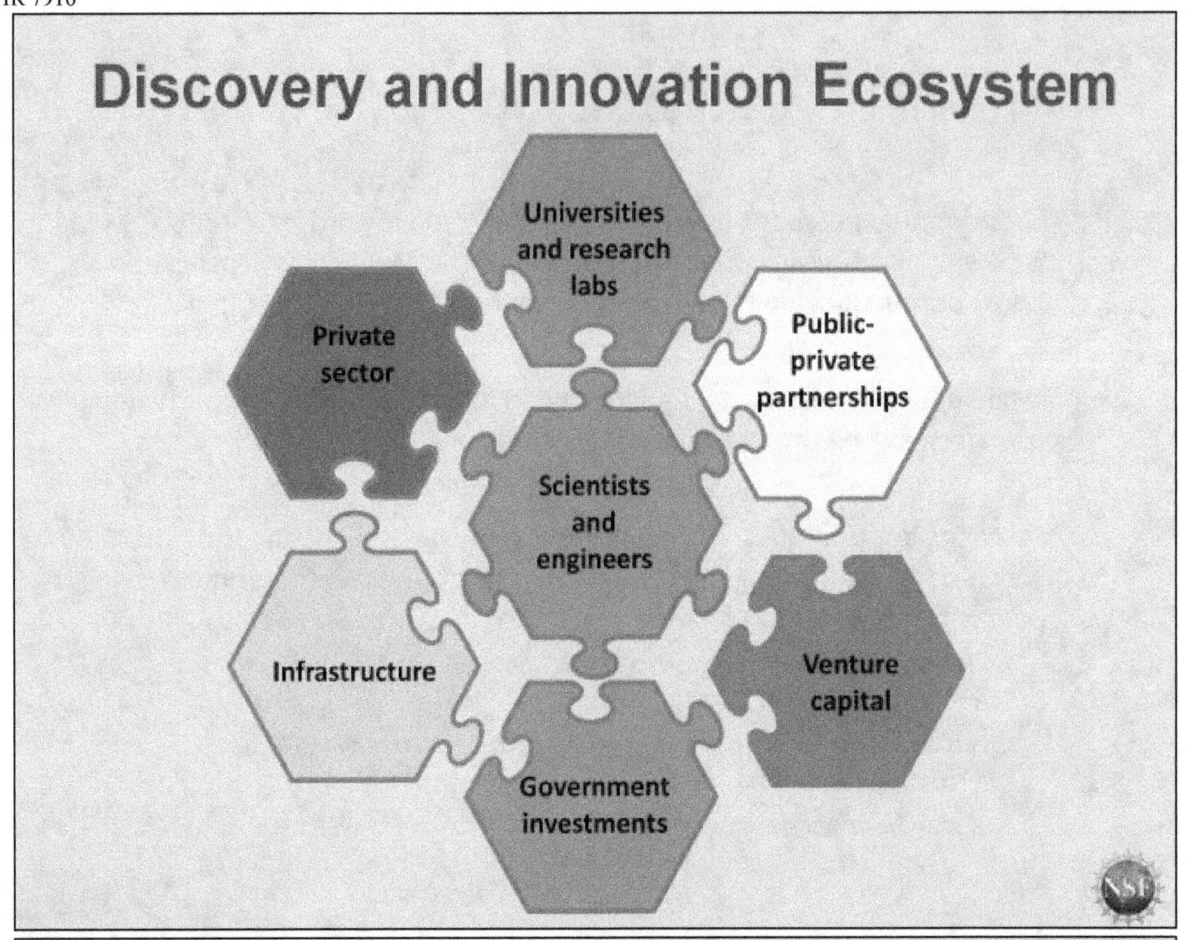

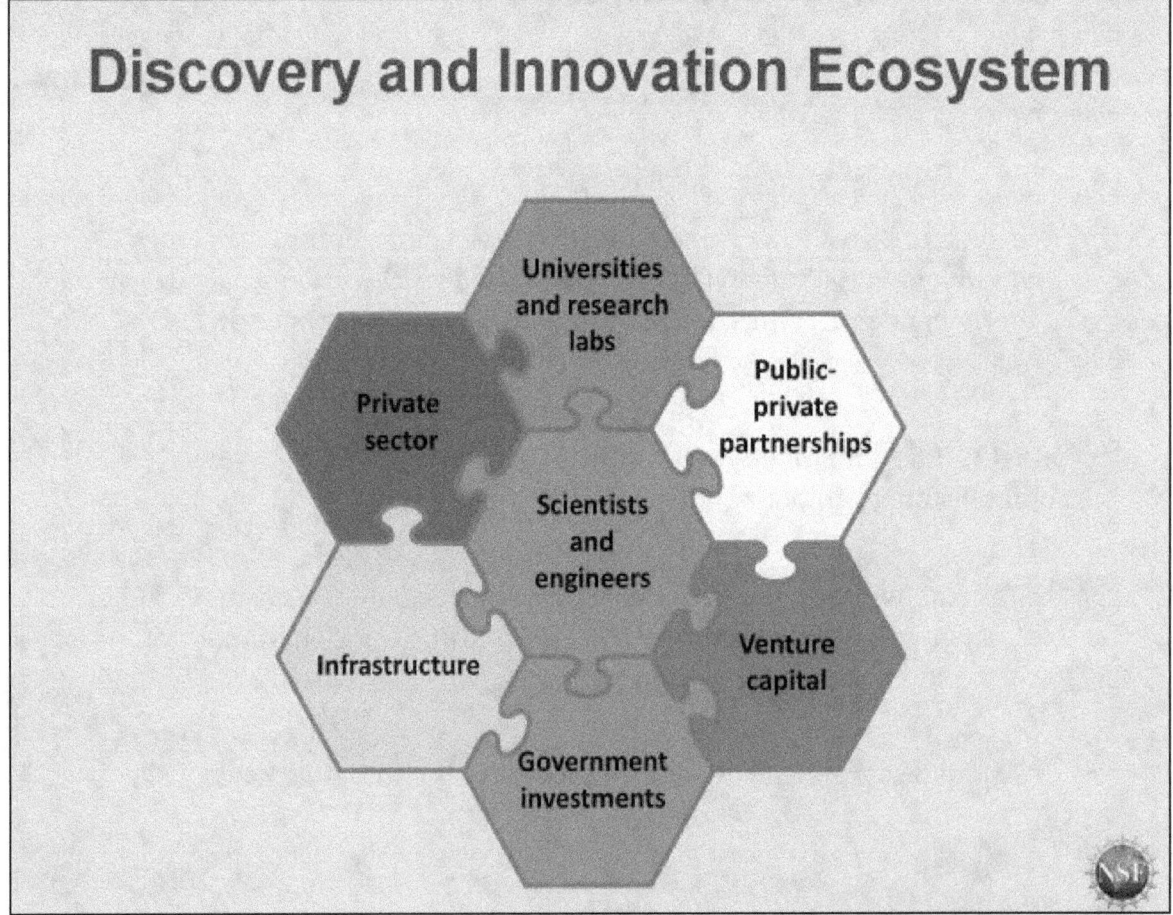

Thanks!

fjahania@nsf.gov

Credits

- Copyrighted material used under Fair Use. If you are the copyright holder and believe your material has been used unfairly, or if you have any suggestions, feedback, or support, please contact: ciseitsupport@nsf.gov.

- Except where otherwise indicated, permission is granted to copy, distribute, and/or modify all images in this document under the terms of the GNU Free Documentation license, Version 1.2 or any later version published by the Free Software Foundation; with no Invariant Sections, no Front-Cover Texts, and no Back-Cover Texts. A copy of the license is included in the section entitled "GNU Free Documentation license" http://commons.wikimedia.org/wiki/Commons:GNU_Free_Documentation_License).

- The inclusion of a logo does not express or imply the endorsement by NSF of the entities' products, services, or enterprises.

6. Security Challenges and Requirements for Control Systems in the Semiconductor Manufacturing Sector

Malek Ben Salem
CyberSecurity Lab
Accenture Technology Labs
Reston, Virginia 20190
Email: malek.ben.salem@accenture.com

I. INTRODUCTION

Moore's Law and the market requirements for higher performance chips are driving the production of increasingly smaller transistors, and therefore, are forcing more stringent controls on semiconductor manufacturing processes and equipment, with a very small room for error.

The second trend in the semiconductor industry is the adoption of the e-Manufacturing paradigm [1]. With the rise of fully-automated factories and the new technology size requirements for chips, new security challenges arise as the control systems are becoming increasingly more complicated. The need for high manufacturing yields using these systems is driving more Advanced Process Controls (APC). Control systems, already ubiquitous in the industry, are becoming more and more sophisticated. And complexity, as is widely acknowledged, is the enemy of security.

The last trend in the industry is the tendency for manufacturers to form joint production ventures. The highly cyclic demand for various consumer electronic products is causing cyclical fluctuation in the manufacturing load of semiconductor factories. The high costs of development and production facilities for different technology node sizes is driving semiconductor companies to form join manufacturing partnerships instead of building new factories. Production of parts may be distributed among manufacturing partner facilities if the manufacturing load in one factory is too high, and part delivery deadlines cannot be met. This new manufacturing model is known the Manufacturing Grid. The goal is to utilize all the manufacturing resources that are distributed between different manufacturing partners and factories different chip parts.

This paper presents threats to controls systems in the semiconductor manufacturing sector that are driven by the above trends in Section II. In Section III, we review recent research work related to the most important threat faced by these control systems. Section IV presents the research priorities and security requirements needed to mitigate these threats. Finally, Section V concludes the paper by summarizing its main points.

II. THREATS AND SECURITY CHALLENGES TO CONTROL SYSTEMS IN THE SEMICONDUCTOR MANUFACTURING SECTOR

We distinguish between targeted attacks to control systems such as Stuxnet [2], and non-targeted attacks. We also distinguish between two types of threats: threats to equipment sensors and controllers, and threats to the IT systems and networks that support these sensors and controllers. For completeness, we cover the major threats of both types in this Section.

A. Equipment Control and Recipe Integrity

Recipes are specifications of equipment processing used to control manufacturing equipment, including processing tool chamber temperature, pressure, and cooling/heating rates. A critical security (and potentially safety) challenge is trusted recipe content, which guarantees that the recipe on the equipment is exactly the one that the factory approved and selected. Another challenge is the traceability of recipe items and parameter usage. A third challenge is preventing DoS attacks, where the adversary prevents the tool controller from receiving recipe parameters and values or sensor measurements by blocking the communication channels between them.

B. Process Data Integrity

The industry applies feed-forward and feedback control, as well as automated fault detection to equipment and to the automated factory, in order to improve process performance and factory yield. These techniques, known as APC rely on the integrity of the data measured by equipment sensors.

Accurate alarm reporting also relies on the accuracy of sensor readings. Alarm reporting is critical to the safety of the equipment, the product, and the factory in general. Alarm reports must be accurate and timely.

One type of targeted attacks against the sensors is the False Data Injection Attack where a malicious third party compromises the integrity of the control systems by controlling the readings of one or more sensors, such as the sensors measuring the ambient temperature inside a chamber on an ion implantation tool.

C. Privilege Over-Entitlement

With the establishment of the Manufacturing Grid, distributed teams are being formed from different companies to collaborate on the development and production of various products. This along with the high job rotation rates among process engineers through various product wafer processing steps, is increasingly complicating the access controls management process. Many engineers quickly accumulate privileges that they do not need to perform their current job functions.

Although this security problem is not strictly related to cyber-physical devices, highly-privileged access to equipment sensors and controllers is a serious threat, knowing that the control systems are increasingly becoming remotely accessible and linked to the corporate networks or to other factories through the Manufacturing Grid. The threat is exacerbated by the open specifications used for process equipment design. This make it easy for malicious users, potentially from a business/manufacturing partner, to launch their attacks and compromise equipment sensors or controllers.

D. Sample Attack

An attacker may develop Stuxnet-like malware [2], featuring zero-day exploits, rootkits, anti-virus evasion techniques, and process injection and hooking code, to target a specific process step within the entire chip manufacturing process.

One of the critical steps in chip manufacturing is the lithography step, where lithography masks or reticles are used to print the pattern of transistors and wires on a microchip. An attacker may substitute a mask with another and use it to print additional transistors and wires on a microchip. Printing as few as a 1000 additional transistors (to the millions of transistors on a chip) may introduce a kill switch or a backdoor to the chip [3]. A kill switch on a chip allows the attacker to stop the chip at any time when he or she sends a specific sequence of bits. All chips on wafers processed using this tool will carry the backdoor or kill switch. The damage may be catastrophic if the chip is installed on a plane for instance. A backdoor may allow the attacker to disable any cryptographic functions that the chip may be running for example. Hardware backdoors create a significant security vulnerability, since hardware is the root of trust, which software builds on.

This is an example of a targeted and very sophisticated attack enabled by the next-generation factory model, where all manufacturing operations are automated and controllers are reachable remotely. The attacker may be able to compromise the controllers of the equipment and have the lithography process tool load the wrong mask. He or she may also compromise the sensors of the tool and/or the software running on the tool so that the wrong process data is reported, thus preventing the detection of the attack through log analysis. The resulting "compromised" chips are hard to detect even during the chip testing phase. Chip makers are not able to test every unspecified function of the device in order to find potential backdoors [3], nor are they able to test all possible sequences of data that might trigger, and therefore discover, a kill switch during testing.

III. RELATED RESEARCH

One of the major attacks against control systems, including process tools used for chip manufacturing, is the *False Data*

Injection Attack. Process tools in this sector are perhaps more vulnerable to these attacks, due to the highly stringent process requirements that have to be met as transistors and technology nodes get smaller. Any false data injected into the control data may drive the process tool "out of control" or irreversibly damage the product.

Mo and Sinopoli studied this type of attack targeting control systems in general. They defined the required and sufficient conditions under which an adversary is able to destabilize a control system that is used to monitor a Linear Time Invariant (LTI) Gaussian system. Linear dynamical systems are one of the most common models for physical systems. The researchers assumed that the control system is equipped with a Kalman filter used to estimate the state of the system from different sensor observations, a LGQ controller used to stabilize the system, and a failure detector [4]. They formulated the action of the attacker as a constrained control problem and showed that, if the attacker knew the variables of the controlled systems and controlled a subset of the sensors, then the attack is feasible. As a defense mechanism, they proposed adding redundant sensors to measure all unstable modes in order to improve the resilience of the control system.

C´ardenas *et al.* reviewed different types of false data injection attacks against control systems, such as bias attacks, surge attacks, and geometric attacks. They experimentally tested these types of attacks against a chemical reactor process. They concluded that it was more important to protect against integrity attacks than DoS attacks. They also found that the proposed data injection attacks could be detected thanks to the slow dynamics of the process. This, probably, does not hold true for semiconductor manufacturing processes.

IV. RESEARCH PRIORITIES AND SECURITY REQUIREMENTS

As we have pointed out, some of the threats targeting control systems are CPS-specific, while others apply to all IT systems. Below, we highlight the security research priorities for control system security in the semiconductor manufacturing sector.

A. Detecting False Data Injection Attacks and Sensor Compromise

The adversary can launch these attacks by obtaining the secret key or by compromising some sensors or controllers. Preventing or at least detecting these attacks is critical. Collecting accurate data is one of the most important conditions for the secure manufacturing of chips. Engineers rely on quality data to make critical decisions related to the availability of manufacturing tools, to the integrity of the specifications of the the manufactured product, and to the reliability and repeatability of the manufacturing process [5].

As the industry embraces the e-Manufacturing model, data integrity becomes even more critical. This requires the protection of the sensor readings and sensor software, eliminating message and data latency and ensuring accurate timestamps. For example, at the equipment controller level, it is important to have accurate readings of the process speed and cooling response rates, the process chamber status, as well as calibration data, and sensor settings. Similarly accurate process data are critical to equipment setup, qualification, process control, and process monitoring.

Data must also be made available in a timely manner to support process control. False data may be an incorrect measurement, an incorrect sender id, or an incorrect timestamp sent with the measurement. Semiconductor processing has stringent timing controls. Distributed factory environments also rely on accurate time in order to coordinate manufacturing processes. So, the time synchronization system used should be fault-tolerant using diverse time sources, so that if one source fails or is inaccessible, others may be reached.

B. Trusted Recipe Management and Fine-Grained Access Control Management

Trusted recipe are a critical security requirement [6]. The management of equipment configuration is of vital importance because configuration changes can cause differences in process capability and outcomes. Trusted recipe management is not a research priority, as much as a security requirement. Security measures that can be used to enforce trusted recipe management are available but cannot easily implemented. Existing access control mechanisms, as previously underlined, do not easily meet the requirements of the industry. New fine-grained access control models to equipment and product recipes are needed in order to help reduce the privilege over-entitlement problem, while allowing design, process, equipment, industrial and integration engineers to solve problems together, especially in cases of manufacturing line emergencies.

C. Dynamic Patching

Control systems are not typically suitable for frequent software patching and updates due to their high availability requirements. Software patches and updates are usually deployed on a fixed, calendar-based schedule, although there is a call to move to condition-based and predictive preventive maintenance [7]. While this may not be the top research priority, we believe that there needs to be more work on dynamic patching for software running on control systems with high uptime requirements.

V. CONCLUSION

In this extended abstract, we gave an overview of the security challenges driven by three trends in the semiconductor manufacturing sector: e-Manufacturing, Moore's Law, and the Manufacturing Grid. We briefly reviewed recent research related to one of the most serious attacks against control systems used by chip makers, namely false data injection attacks. Then, we highlighted the security-related research priorities for the industry.

Control systems are subject to non-targeted attacks and targeted attacks, such as Stuxnet [2]. Attacks against control systems may be directed at the sensors of the control systems, their actuators or controllers, or the IT systems and networks supporting the information processing and communication.

Techniques for detecting tampering, and validating the inputs provided by the sensors are of paramount importance.

Besides their unique security requirements, control systems share many of the security requirements with traditional IT systems. However, if we are to achieve secure control systems in this sector, we need to model the security implications of the physical interactions in semiconductor processing tools. Design of the equipment, including hardware components and the software that it runs, as well as the factory automation platform, should consider security as part of system architecture and software development. Information flow and control paths have to be identified during the design phase, so that the operational security requirements of the system may be met.

REFERENCES

[1] T. Walker and D. Stark, "Next-generation factory publications by the semiconductor manufacturing consortium (sematech)." [Online]. Available: http://sematech.org/docubase/wrappers/33s.htm

[2] N. Falliere, L. O. Murch, and E. Chien, "W32.stuxnet dossier version 1.4," New York, NY, USA, February 2011.

[3] S. Adee, "The hunt for the kill switch," Spectrum, IEEE, vol. 45, no. 5, pp. 34–39, may 2008.

[4] Y. Mo and B. Sinopoli, "False data injection attacks in control systems," in Proceedings of the First Workshop on Secure Control Systems, ser. SCS '10. Team for Research in Ubiquitous Secure Technology (TRUST), April 2010. [Online]. Available: http://www.cse.psu.edu/smclaugh/cse598e-f11/papers/mo.pdf

[5] G. Crispieri, "Data quality guidelines: Version 1," July 2008. [Online]. Available: http://sematech.org/docubase/document/4843ceng.pdf

[6] L. Rist, "Recipe management system: Common requirements and benchmark 2007," May 2008. [Online]. Available: http://sematech.org/docubase/document/4930atr.pdf

[7] T. Walker and D. Stark, "Ismi predictive preventive maintenance implementation guideline," October 2010. [Online]. Available: http://sematech.org/docubase/document/5119atr.pdf

Malek Ben Salem, PhD, CISSP Dr. Ben Salem has an extensive background in process controls. She has worked for 9 years as an IT systems developer and integrator and as a Controls engineer at IBM's East Fishkill chip manufacturing factory. Dr. Ben Salem earned a PhD degree in Computer Science from Columbia University in 2012. Her research interests include intrusion detection, network security, and secure dependable systems.

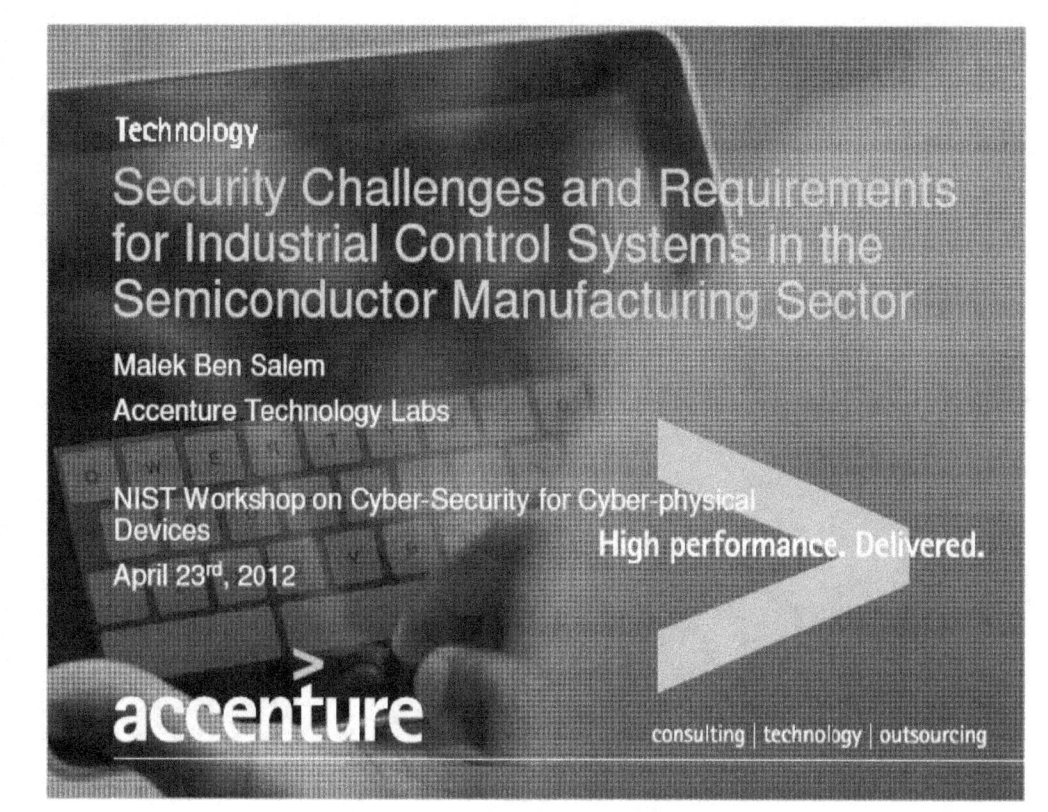

Technology

Security Challenges and Requirements for Industrial Control Systems in the Semiconductor Manufacturing Sector

Malek Ben Salem

Accenture Technology Labs

NIST Workshop on Cyber-Security for Cyber-physical Devices
April 23rd, 2012

High performance. Delivered.

accenture

consulting | technology | outsourcing

Outline

- Background Information
- Security Challenges
- Sample Attack
  - Insertion of Hardware Trojans
  - Failure of Existing Common Hardware Trojan Detection Approaches
- Research Priorities
- Summary

Semiconductor Manufacturing: Background Information

Chip Manufacturing Process Overview

Semiconductor device fabrication is a series of four types of processing steps: deposition, etching, patterning, and modification of electrical properties. Additional measurement/metrology steps are added.

 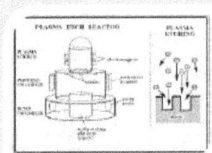  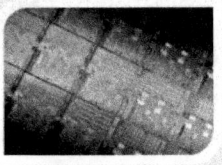

Deposition

Growing /transferring material onto wafer, wafer coating .

E.g. Wafers are put into a copper sulphate solution, and Copper ions are deposited onto the transistor through a process called electroplating.

Etching

Removing material from the wafer either in bulk or selectively process used between levels.

E.g. Chemical Mechanical Planarization (CMP)

Lithography

Patterning and shaping of wafer materials

E.g. wafer costing with a photo-resist that gets exposed by a stepper, a machine that focuses, aligns, and moves the mask exposing select portions of the wafer to short wavelength light.

Electrical Property Modification

Doping transistor sources and drains by diffusion furnaces and by ion implantation

Activating implanted dopants through Furnace or Rapid Thermal Anneal (RTA)

Pictures courtesy of spectrum.ieee.org, intel.com, and poli.cs.vsb.cz.

3

Trends in Semiconductor Manufacturing

- Moore's Law and the market requirements for higher performance chips are driving the production of smaller transistors
  - Smaller devices and larger wafers
- Adoption of the e-Manufacturing paradigm
  - Fully-automated factories
- Control systems are more complicated
- Tighter tolerance windows
- More stringent process controls are implemented on semiconductor manufacturing processes and equipment

4

Trends in Semiconductor Manufacturing (contd.)

- Economic and market forces drive outsourcing IC fabrication
  - Compromising the IC supply chain for sensitive commercial and defense applications becomes easy.
  - Attacker could substitute Trojan ICs for genuine ICs during transit.
  - Attacker could subvert the fabrication process itself by implanting additional Trojan circuitry into the IC mask.
- Manufacturing Grid: Joint production platforms
  - Cyclic demand for consumer electronic products
  - High costs of development and production facilities for different technology node and wafer sizes
  - Load distribution among manufacturing partner facilities
- Objectives:
  - Optimize all the distributed manufacturing resources
  - Minimize IP disclosure

5

NISTIR 7916

Security-Related Challenges

6

Equipment Control and Recipe Integrity

- Recipes:
  - Specifications of equipment processing
  - Used to control manufacturing equipment, including processing tool chamber temperature, pressure, and cooling/heating rates.

- Critical Security Issues
  - Trusted recipe content to ensure that the recipe on the equipment is exactly the one that the factory approved and selected.
  - Traceability of recipe items and parameter usage
  - Preventing DoS attacks and blocking the communication channels between equipment controllers and sensors or recipe databases

Picture courtesy of seconsemi.com

7

78

Process Data Integrity

- Advanced Process Controls (APC) are critical for high-quality process performance and factory yield
  - Feed-forward and feedback control
  - Automated fault detection to equipment and to the automated factory, in order to improve process performance and factory yield.

- These techniques, known as APC rely on the integrity of the data measured by equipment sensors.
  - Accurate sensor readings
  - Accurate and timely alarm reporting
  - Alarm reporting is critical to the safety of the equipment, the product, and the factory in general.

8

False Data Injection Attacks

- Malicious third party compromises the integrity of the control systems by controlling the readings of one or more sensors
  - e.g. sensors measuring the ambient temperature inside a chamber on an Ion implantation tool

- APC is vulnerable to false data injection attacks.
  - Consequence: scrapped wafers

- High scrap costs
  - Average wafer cost ~$9000 (depending on product and process step)
  - Wafers are processed in lots of 25 wafers
  - MWTD (Mean-Wafers-To-Detect) depends on sampling plan and process performance.

Picture courtesy of rubbertechnology.info

9

Privilege Over-Entitlement

- High job rotation rates
  - Process engineers rotate through various product wafer processing steps
  - Engineers rotate between design, process and integration roles
  - Complicated access controls management to product and equipment recipes

- Many engineers quickly accumulate privileges that they do not need to perform their current job functions.

- Highly-privileged access to equipment sensors and controllers is a serious threat
  - Serious problem, although not strictly related to cyber-physical devices
  - Exacerbated by remotely accessible control system, distributed global teams, and open specifications used for process equipment design.

10

Sample Attack:
Hardware Trojans

11

Hardware Trojans in the News

Dell warns of hardware Trojan

Computer maker Dell is warning that some of its server motherboards have been delivered to customers carrying an unwanted extra: computer malware. It could be confirmation that the "hardware Trojans" ... are indeed a real threat .

- Homeland Security News Wire July 2010

F.B.I. Says the Military Had Bogus Computer Gear

...the .. sinister specter of an electronic Trojan horse, lurking in the circuitry of a computer or a network router and allowing attackers clandestine access or control, was raised .. by the FBI and the Pentagon.
The new law enforcement and national security concerns were prompted by Operation CISCO Raider, which has led to 15 criminal cases involving counterfeit products bought in part by military agencies, military contractors and electric power companies in the United States.

-The New York Times, May 2008

12

Hardware Trojans

- Monitor for a specific but rare trigger condition
  - e.g., a specific bit pattern in received data packet or on a bus
  - until a timer reaches a particular value.

- Hardware is the root of trust
  - Software security mechanisms can be bypassed by malicious hardware.

- Potential targets
  - Hardware used for defense
  - Commercial grade cryptographic and security critical hardware

- Look genuine ICs with normal input/output behavior during testing and normal use.

- Tampering is very difficult to detect and mitigate
  - Hard to detect using visual inspection or conventional testing techniques

13

Hardware Trojans

- Trojans may be inserted during the design or manufacturing
  - Long supply chain
  - Complexity increases vulnerability
- Capable of inflicting catastrophic damage
  - Modify chip's function through additional logic or by removing or bypassing existing logic
    - Disabling encryption
    - Clock disruption to shut down the chip or affect its synchronization
    - Adding glitches to compromise system integrity and security (backdoor)
    - Destruction of the operating environment of original circuit
      - Shutting down power (kill-switch), generating noise to disrupt critical signals, or increasing thermal gradients on the chip possibly causing burn out
  - Modify chip's parametric properties
    - E.g. delay by modifying wire and transistor geometries

14

Photolithography

- Process used to remove parts of a thin-film or substrate
- Uses light to transfer a geometric pattern from a photomask
- Includes several steps
  - Wafer Cleaning, Barrier Formation and Photoresist Application
  - Soft-Baking
  - Mask Exposure
  - Printing
  - Development
  - Hard-Baking

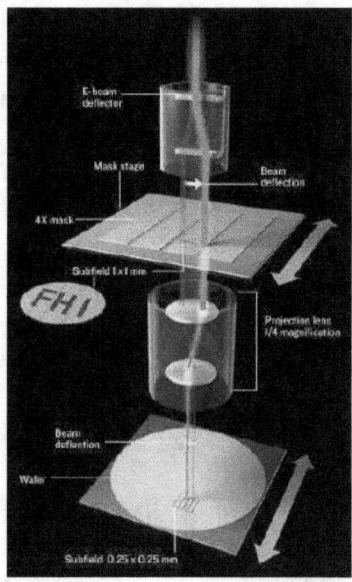

15

Conventional Multi-layer Lithography : Stepping

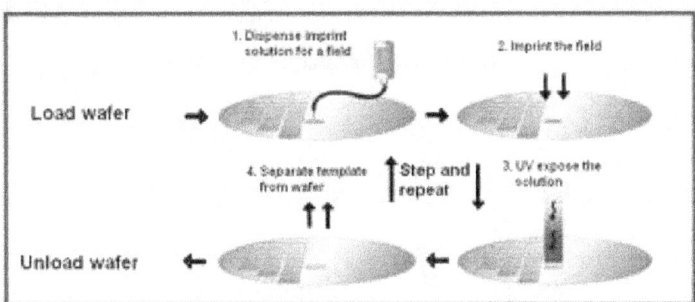

- Composed of one patterning step and several steps of oriented deposition
- Most lithographic techniques are 2-dimensional (photolithography, e-beam lithography, and imprint lithography)
- Using the wrong mask affects all dies on a wafer
- All chemicals are loaded automatically into the tool, and controlled by recipe items.

16

Many Opportunities for Malicious Insiders

Lithography processes present opportunities to print additional circuitry and devices

Need to replace glass masks
Masks are automatically loaded into litho tools
No physical access to target tool required

Other process and measurement/metrology steps present opportunities for causing scraps

Silicon Wafer | Manufacturing Line | Functional Dies

Trojan circuitry may be inserted in different layers of circuitry within the chip

Long manufacturing lines ~200 processing steps

Many opportunities for malicious insiders

Targeting processes at the BEOL (Back End Of the Line) causes higher damages to the IC manufacturer.

17

Transistors Formed from a Single Lithography Step [3]

- Topographically Encoded Micro-Lithography (TEMIL)
  - Single level of topography (photolithography or molding)
  - A substrate with multiple shadow evaporations
- Shadow Evaporation
  - All information needed to fabricate complex structures is encoded in the topography of patterned polymer
- May replace several steps of lithography
  - One lithography step
  - Sequential shadow evaporation/deposition steps of various materials
  - Each functional layer of device can be deposited independently using a single level of topography
- Produce transistors without any doping, etching, or lithography alignment steps
- Malicious insider needs access to one tool/recipe only

18

Hardware Trojan Activation

- Trigger Type
  - Ticking time-bomb triggers: Open to everyone
  - Data triggers: Hacker needs access to the machine to trigger
- Externally-activated
  - Using a receiver or antenna on chip
  - Forcing internal registers to specific date to extract secret keys
- Internally-activated
  - Always-on: Trojan continuously active, implemented by modifying the geometries of the chips, such that certain nodes or paths in the chip have a higher susceptibility to failure (parametric Trojans)
  - Condition-based (Temperature, pressure, or voltage sensor output / Internal logic state / Input pattern / Internal counter value.
    - Implemented by adding logic gates and/or flip-flops to the chip)
    - Represented as a combinational or sequential circuit

19

Failure of Existing Common Solutions

- Currently impossible to certify the trustworthiness of processors & controllers as Trojan detection is very hard
- Nano-scale devices and high system complexity make detection through physical inspection almost impossible.
- Inspection through destructive reverse engineering does not guarantee absence of Trojans in ICs not destructively inspected.
- Audits not very effective at catching bugs
- Obfuscation during fabrication
  - Motivated attacker can always identify criticality of manufactured IC
  - Shown to be impossible to achieve in most cases
- Triggers are finite state machines that can change states when **time** or **input data** changes

20

Trojan Detection: Failure Analysis

- Techniques
  - Scanning optical microscopy (SOM)
  - Scanning electron microscopy (SEM)
  - Pico-second imaging circuit analysis (PICA)
  - Voltage contrast imaging (VCI)
  - Light-induced voltage alternation (LIVA)
  - Voltage alternation CIVA
- Effective, but expensive and time-consuming
- Require destructively using at least one sample chip
- Many ineffective for technologies in the nano-meter domain
- Not effective for randomly inserted Trojans

21

Trojan Detection: ATPG (Automatic Test Pattern Generation)

- Uses standard VLSI fault detection tools
- Applies a digital stimulus and inspects digital output of chip
- Digital stimulus is derived using the netlist of the chip
  - For parametric Trojans of the parametric type, the netlist of a chip is the same with and without the Trojan
- Likely to yield best results of parametric Trojans
  - Due to stealthy activation criteria
  - ATPG directed to generate tests for nodes and paths that are hard-to-detect (i.e., difficult to control and/or observe,)
- Not effective with functional Trojans
  - Trigger condition occurs with very low probability during functional testing
  - $1/2^{64}$ probability of getting detected during validation

22

Trojan Detection: Side Channel Analysis

- Effective in extracting information about internal operations of embedded devices
  - Timing, Power consumption, Electromagnetic emanation profiles
  - Differential Power, Electromagnetic (EM) Analysis
  - Average measurements from multiple samples to deal with noise problem
- Approach
  - Requires destruction of a few ICs to validate authenticity
  - Other ICs validated using side-channel analysis for absence of any significantly sized Trojans (3-4 orders of magnitude smaller than IC [2])
- Effective for detection of functional Trojans
  - Detects functional Trojans without activating them, i.e., through the measurement of their secondary action characteristics
  - Not effective for testing circuits at extremely low clock frequencies

23

Research Priorities and Security Requirements

24

Accurate Data Collection

- Accurate data is critical to secure chip manufacturing
  - Equipment availability decision
  - Integrity of the specifications of the manufactured product
  - Reliability and repeatability of the manufacturing process [5]

- Data integrity becomes more important with the adoption of the e-Manufacturing model
  - E.g. accurate readings of the process speed and cooling response rates, the process chamber status, calibration data, and sensor settings at the equipment controller level

- Accurate process data are critical to equipment setup, qualification, process control, and process monitoring.

- Data collection timeliness needed to support process control

25

Preventing/Detecting False Data Injection Attacks and Sensor Compromise

- Attacks possible through sensor compromise or by obtaining the secret key
- Preventing/detecting these attacks is critical.
- This requires the protection of the sensor readings and sensor software, eliminating message and data latency and ensuring accurate timestamps.
- Fault-tolerant time synchronization system using diverse time sources.

26

Trusted Recipe Management

- Trusted recipes are a critical security requirement
  - Trusted management of equipment configuration
  - Configuration changes can cause differences in process capability and outcomes
- Security measures to enforce trusted recipe management are needed
- Existing access control mechanisms do not meet the requirements of the industry
  - Equipment engineers with administrator privileges

27

Fine-Grained Access Control Management

- New fine-grained access control models to equipment and product recipes are needed
- Need to reduce the privilege over-entitlement problem
  - Allowing design, process, equipment, industrial and integration engineers to solve problems together,
  - Consider manufacturing line emergencies

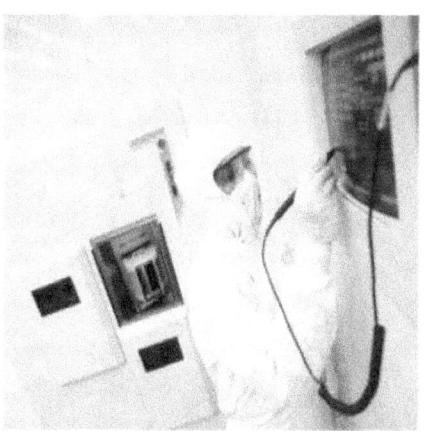

A technician programs a product "Recipe" into an epi reactor.

28

Dynamic Patching

- Control systems are not typically suitable for frequent software patching and updates due to their high availability requirements.
- Software patches and updates are usually deployed on a fixed, calendar-based schedule
- Call to move to condition-based and predictive preventive maintenance .

29

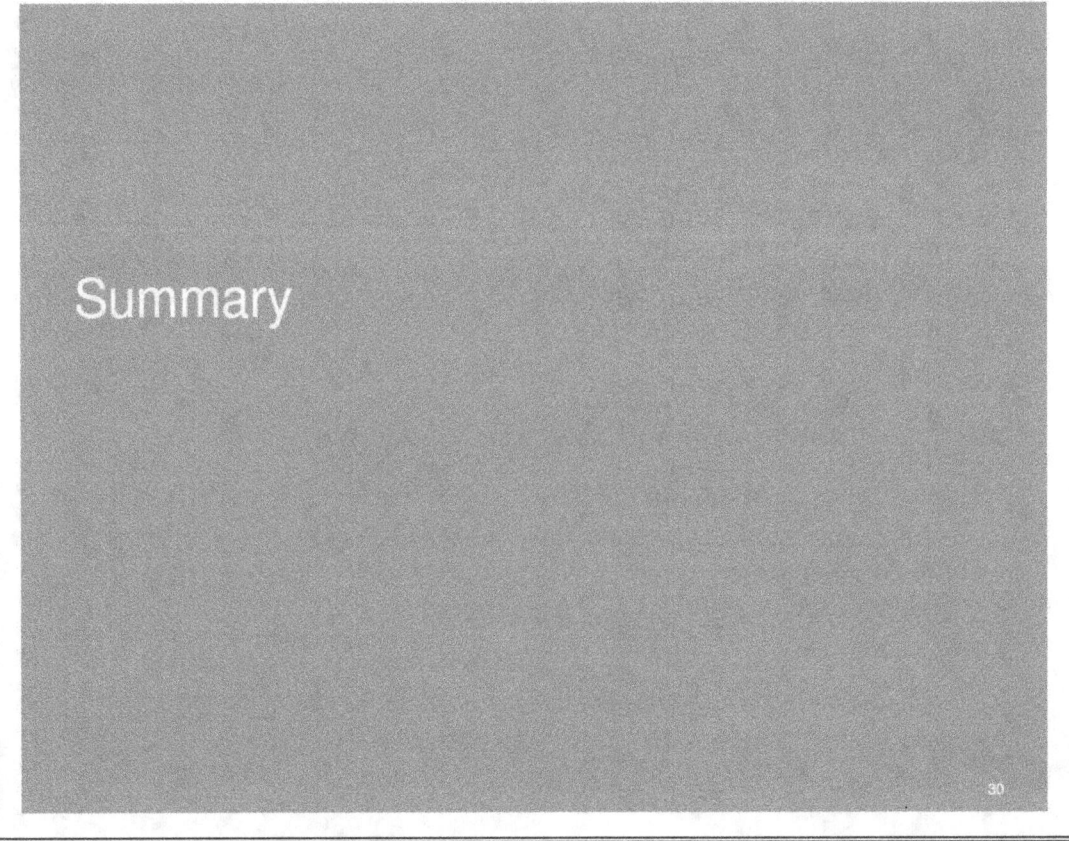

Summary

30

Threats and Security Challenges in the Semiconductor Manufacturing Sector

- Threats
  - Threats to IT systems and networks
  - Threats to equipment sensors and controllers
- Attacks
  - Regular attacks
  - Targeted attacks
    - Process vs. final product
    - Sabotage vs. espionage
- Security Challenges
  - Equipment Control and Recipe Integrity
  - Process Data Integrity
  - Privilege Over-Entitlement

31

Conclusion

- Existing Hardware Trojan detection techniques not very effective
  - Detection during manufacturing may be more effective
  - Mask signatures
- Need to model the security implications of the physical interactions in semiconductor processing tools
- Need to consider security as part of system architecture and software development for
  - Semiconductor processing and measurement/metrology tools
    - Information flow and control paths have to be identified
    - Joint work between IC and tool manufacturing companies
  - Plant automation infrastructure

32

References

[1] S. Adee "The Hunt for the Kill Switch", IEEE Spectrum Vol. 45 Num. 5, pp 34-39, May 2008.
http://spectrum.ieee.org/semiconductors/design/the-hunt-for-the-kill-switch/0

[2] D. Agrawal, S. Baktir, D. Karakoyunlu, P. Rohatgi, and B. Sunar. "Trojan detection using IC fingerprinting." Proceedings of the 28th IEEE Symposium on Security & Privacy, pp 296–310, May 2007.

[3] M. D. Dickey, K. J. Russell, D. J. Lipomi, V. Narayanamurty, and G. M. Whitesides. "Transistors Formed from a Single Lithography Step Using Information Encoded in Topography" - 2010 Wiley-VCH Verlag GmbH & Co. KGaA, Weinheim. http://www.small-journal.com.

[4] M. Hicks, M. Finnicum, S. King, M. Martin, and J. Smith. "Overcoming an Untrusted Computing Base: Detecting and Removing Malicious Hardware Automatically." Proceedings of the 31st IEEE Symposium on Security & Privacy, pp 159–172, May 2010.

33

References

[5] S. King, J. Tueck, A. Cozzie, C. Grier. W. Jiang, and Y. Zhou. "Designing and implementing malicious hardware." Proceedings of the 1st Usenix Workshop on Large-Scale Exploits and Emergent Threats, pp 1-8, 2008.

[6] SEMATECH technical Publications. http://www.sematech.org/publications/technical.htm

[7] C. Sturton, M. Hicks, D. Wagner, and S. T. King. "Defeating UCI: Building Stealthy and Malicious Hardware." Proceedings of the 32nd IEEE Symposium on Security & Privacy, pp 64-77, 2011.

[8] X. Wang and M. Tehranipoor. "Detecting Malicious Inclusions in Secure Hardware: Challenges and Solutions." Proceedings of the 2008 IEEE International Workshop on Hardware-Oriented Security and Trust, pp 15–19, 2008.

[9] A. Waskman and S. Sethumadhavan. "Tamper Evident Microprocessors". Proceedings of the 31st IEEE Symposium on Security and Privacy, 2010.

34

References

[10] A. Waskman and S. Sethumadhavan. "Silencing Hardware Backdoors." Proceedings of the 32nd IEEE Symposium on Security & Privacy, pp 49-63, 2011.

[11]"Dell warns of hardware trojan." Homeland Security News Wire, July 2010. http://www.homelandsecuritynewswire.com/dell-warns-hardware-trojan

[12] J. Markoff. "F.B.I. Says the Military Had Bogus Computer Gear." The New York Times, May 2008. http://www.nytimes.com/2008/05/09/technology/09cisco.html?_r=3&partner=rssnyt&emc=rss

35

7. Keynote Speaker: Tom Dion

Tom Dion
Vendor Assessment Lead
Control Systems Security Program
Department of Homeland Security

Control Systems Security Program (CSSP)

The goal of the DHS National Cyber Security Division's CSSP is to reduce industrial control system risks within and across all critical infrastructure and key resource sectors by coordinating efforts among federal, state, local, and tribal governments, as well as industrial control systems owners, operators and vendors. The CSSP coordinates activities to reduce the likelihood of success and severity of impact of a cyber attack against critical infrastructure control systems through risk-mitigation activities.

Cyber Security Evaluation Tool

Critical infrastructures are dependent on information technology systems and computer networks for essential operations. Particular emphasis is placed on the reliability and resiliency of the systems that comprise and interconnect these infrastructures. NCSD collaborates with partners from across public, private, and international communities to advance this goal by developing and implementing coordinated security measures to protect against cyber threats.

The Cyber Security Evaluation Tool (CSET™) is a Department of Homeland Security (DHS) product that assists organizations in protecting their key national cyber assets. It was developed under the direction of the DHS National Cyber Security Division (NCSD) by cybersecurity experts and with assistance from the National Institute of Standards and Technology. This tool provides users with a systematic and repeatable approach for assessing the security posture of their cyber systems and networks. It includes both high-level and detailed questions related to all industrial control and IT systems.

CSET is a desktop software tool that guides users through a step-by-step process to assess their control system and information technology network security practices against recognized industry standards. The output from CSET is a prioritized list of recommendations for improving the cybersecurity posture of the organization's enterprise and industrial control cyber systems. The tool derives the recommendations from a database of cybersecurity standards, guidelines, and practices. Each recommendation is linked to a set of actions that can be applied to enhance cybersecurity controls.

CSET has been designed for easy installation and use on a stand-alone laptop or workstation. It incorporates a variety of available standards from organizations such as National Institute of Standards and Technology (NIST), North American Electric Reliability Corporation (NERC), International Organization for Standardization (ISO), U.S. Department of Defense (DoD), and others. When the tool user selects one or more of the standards, CSET will open a set of questions to be answered. The answers to these questions will be compared against a selected security assurance level, and a detailed report will be generated to show areas for potential improvement. CSET provides an excellent means to perform a self-assessment of the security posture of your control system environment.

For more information about the DHS Control Systems Security Program, please visit:
http://www.us-cert.gov/control_systems/

For more information about the CSET, please visit:
http://www.us-cert.gov/control_systems/satool.html

NISTIR 7916

8. Application of Dynamic System Models and State Estimation Technology to the Cyber Security of Physical Systems

Barry Horowitz, Kate Pierce
Systems and Information Engineering
University of Virginia Charlottesville, VA, USA
Email: bh8e@virginia.edu, kmp7ef@virginia.edu

As advances in technology permit automatic control of more and more of the functions of physical systems, the opportunity for cyber attacks that include exploitation of such automation capabilities becomes a greater risk. For example, in the 2010 Stuxnet attack an embedded infection in control systems was used to successfully damage a large number of nuclear power related centrifuges in Iran. While the application of perimeter security technologies have been applied to help manage the likelihood of SCADA-based cyber attackers exploiting highly automated physical systems, successful attacks have occurred, and furthermore, perimeter solutions do not address important classes of insider and supply chain initiated attacks. As a result, it has been recognized that perimeter security needs to be augmented by other approaches for addressing potential cyber attacks [1].

Frequently, as a means for added operational assurance, highly automated physical systems include the presentation of system status information that permits human operators to take controlling actions when the automated system appears to be operating in an out-of-normal manner. For example, the operation of a turbine may be automatically controlled, but operators can observe critical information regarding the turbines operation, such as vibration levels, temperature, and rotation rate. If the operator observes measurements that are outside the designated region of proper operation, specific manual actions can be required of the operator in order to avoid undesirable consequences [2]. However, as was the case in the Stuxnet attacks, the cyber attacker can not only manipulate a physical systems performance through infections in its control system, but can also manipulate data presented to operators; data that can, when utilized within standard operating procedures, either stimulate inappropriate control actions or prevent needed control actions on an operators part. In the case of the turbine example, a successful cyber attack can result in indications to operators that would imply that all is well when it is not, or indications that would call for disruptive operator action when, in reality, none is required (e.g., unnecessarily shutting down the turbine). Note that it is quite typical for operator displays to be designed for simplicity, so that critical manual actions will not be delayed by human limitations related to viewing and interpreting too much information. As a result, physical systems typically include measurement and collection of information that could conceivably be used, but is not, for automation override decision making. For example, driving an automobile involves a driver monitoring a few of the many available engine state measurements that could be made available for viewing, but could confuse the driver while offering little, if any, benefit.

This paper presents an approach for addressing cyber attacks on physical systems that include purposeful manipulation of operator displays. The presented approach involves embedding security functions within the physical system being protected; functions that can be the basis for detection of inconsistent system dynamics data derived from measurements within the system that is being protected. In particular, the use of dynamic mathematical models of physical systems in combination with state estimation techniques is suggested as the basis for system architectures that can be employed to detect situations where information displays for system operators are being manipulated as part of a cyber attack. One can divide the states of a physical system into 3 classes: 1) those that are presented to operators for control purposes, but are considered as least trusted from a cyber security viewpoint; 2) those that can be measured and analyzed in segregated equipment from the equipment being used for measuring, analyzing and displaying of least trusted states, but are not used for operator assistance and are considered as more trusted; and 3) those that are not measured. The paper draws on dynamic state estimation techniques to develop, when feasible, estimates of the least trusted states values, and the variances of these estimates, from measurements of more trusted states. Systems that satisfy control system conditions for being observable satisfy the sufficient conditions for this cyber security solution. The paper shows how these estimates can provide the basis for detecting attacks on automatic control systems that include manipulation of data presented to operators, and how this approach can be used to manage system restoration. Theoretical results are presented for a range of physical system models, and a specific system model for an electrical generator is used to illustrate the performance one can achieve in an actual application, including calibration of expected performance in terms of missed detections, false detections and delay time for detections of cyber attacks.

This material is based upon work supported in part by the U.S. Department of Defense through the Systems Engineering Research Center (SERC) under Contract H98230-08-D-0171. SERC is a federally funded University Affiliated Research Center managed by Stevens Institute of Technology.

REFERENCES

[1] W. A. Wulf and A. K. Jones, "Reflections on cyber security," *Science Magazine*, vol. 326, pp. 943–944, 2009.

[2] R. A. Jones, T. V. Nguyen, and B. M. Horowitz, "System-aware security for nuclear power systems," in *2011 IEEE International Conference on Technologies for Homeland Security*, November 2011.

System Aware Cyber Security

Application of Dynamic System Models and State Estimation Technology to the Cyber Security of Physical Systems

Barry M. Horowitz, Kate Pierce

University of Virginia

April, 2012

This material is based upon work supported, in whole or in part, by the U.S. Department of Defense through the Systems Engineering Research Center (SERC) under Contract H98230-08-D-0171. SERC is a federally funded University Affiliated Research Center managed by Stevens Institute of Technology

Objectives for System Aware Cyber Security Research

- Increase cyber security by developing new system engineering-based technology that provides a Point Defense option for cyber security
  - Inside the system being protected, for the most critical functions
  - Complements current defense approaches of network and perimeter cyber security
- Directly address supply chain and insider threats that perimeter security does not protect against
  - Including physical systems as well as information systems
- Provide technology design patterns that are reusable and address the assurance of data integrity and rapid forensics, as well as denial of service
- Develop a systems engineering scoring framework for evaluating cyber security architectures and what they protect, to arrive at the most cost-effective integrated solution

Publications

Jennifer L. Bayuk and Barry M. Horowitz, An Architectural Systems Engineering Methodology for Addressing Cyber Security, Systems Engineering 14 (2011), 294-304.

- Rick A. Jones and Barry M. Horowitz, System-Aware Cyber Security, ITNG, 2011 Eighth IEEE International Conference on Information Technology: New Generations, April, 2011, pp. 914-917. (Best Student Paper Award)
- Rick A. Jones and Barry M. Horowitz, System-Aware Security for Nuclear Power Systems, 2011 IEEE International Conference on Technologies for Homeland Security, November, 2011. (Featured Conference Paper)
- Rick A. Jones and Barry M. Horowitz, A System-Aware Cyber Security Architecture, Systems Engineering, Volume 15, No. 2, February, 2012

System-Aware Cyber Security Architecture

- System-Aware Cyber Security Architectures combine design techniques from 3 communities
  - Cyber Security
  - Fault-Tolerant Systems
  - Automatic Control Systems
- The point defense solution designers need to come from the communities related to system design, providing a new orientation to complement the established approaches of the information assurance community
- New point defense solutions will have independent failure modes from traditional solutions, thereby minimizing probabilities of successful attack via greater defense in depth

A Set of Techniques Utilized in System-Aware Security

Cyber Security	Fault-Tolerance	Automatic Control
*Data Provenance	*Diverse Redundancy	*Physical Control for
*Moving Target	(DoS, Automated Restoral)	Configuration Hopping
(Virtual Control for Hopping)	*Redundant Component Voting	(Moving Target, Restoral)
*Forensics	(Data Integrity, Restoral)	*State Estimation
		(Data Integrity)
		*System Identification
		(Tactical Forensics, Restoral)

A Set of Techniques Utilized in System-Aware Security

Cyber Security	Fault-Tolerance	Automatic Control
*Data Provenance	*Diverse Redundancy	*Physical Control for
*Moving Target	(DoS, Automated Restoral)	Configuration Hopping
(Virtual Control for Hopping)	*Redundant Component Voting	(Moving Target, Restoral)
*Forensics	(Data Integrity, Restoral)	*State Estimation
		(Data Integrity)
		*System Identification
		(Tactical Forensics, Restoral)

This combination of solutions requires adversaries to:
- Understand the details of how the targeted systems actually work

A Set of Techniques Utilized in System-Aware Security

Cyber Security	Fault-Tolerance	Automatic Control
*Data Provenance	*Diverse Redundancy	*Physical Control for
*Moving Target	(DoS, Automated Restoral)	Configuration Hopping
(Virtual Control for Hopping)	*Redundant Component Voting	(Moving Target, Restoral)
*Forensics	(Data Integrity, Restoral)	*State Estimation
		(Data Integrity)
		*System Identification
		(Tactical Forensics, Restoral)

This combination of solutions requires adversaries to:
- Understand the details of how the targeted systems actually work
- Develop synchronized, distributed exploits consistent with how the attacked system actually works

A Set of Techniques Utilized in System-Aware Security

Cyber Security	Fault-Tolerance	Automatic Control
*Data Provenance	*Diverse Redundancy	*Physical Control for
*Moving Target	(DoS, Automated Restoral)	Configuration Hopping
(Virtual Control for Hopping)	*Redundant Component Voting	(Moving Target, Restoral)
*Forensics	(Data Integrity, Restoral)	*State Estimation
		(Data Integrity)
		*System Identification
		(Tactical Forensics, Restoral)

If implemented properly, this combination of solutions requires adversaries to:
- Understand the details of how the targeted systems actually work
- Develop synchronized, distributed exploits consistent with how the attacked system actually works
- Corrupt multiple supply chains

Example Design Patterns Under Development

- **<u>Diverse Redundancy</u>** for post-attack restoration
- **<u>Diverse Redundancy</u> + <u>Verifiable Voting</u>** for trans-attack defense
- **<u>Physical Configuration Hopping</u>** for moving target defense
- **<u>Virtual Configuration Hopping</u>** for moving target defense
- **<u>Physical Confirmations of Digital Data</u>**
- <u>Data Consistency Checking</u>

ATTACK 1: OPERATOR DISPLAY ATTACK

ATTACK 2: CONTROL SYSTEM & OPERATOR DISPLAY ATTACK

ATTACK 3: SENSOR SYSTEM ATTACK

ATTACKS 1 & 2
OPERATOR DISPLAY ATTACK/
COORDINATED CONTROL SYSTEM &
OPERATOR DISPLAY ATTACK

The Problem Being Addressed

- Highly automated physical system
- Operator monitoring function, including criteria for human over-ride of the automation
- Critical system states for both operator observation and feedback control – consider as *least trusted from cyber security viewpoint*
- Other measured system states – consider as *more trusted from cyber security viewpoint*
- CYBER ATTACK: Create a problematic outcome by disrupting human display data and/or critical feedback control data.

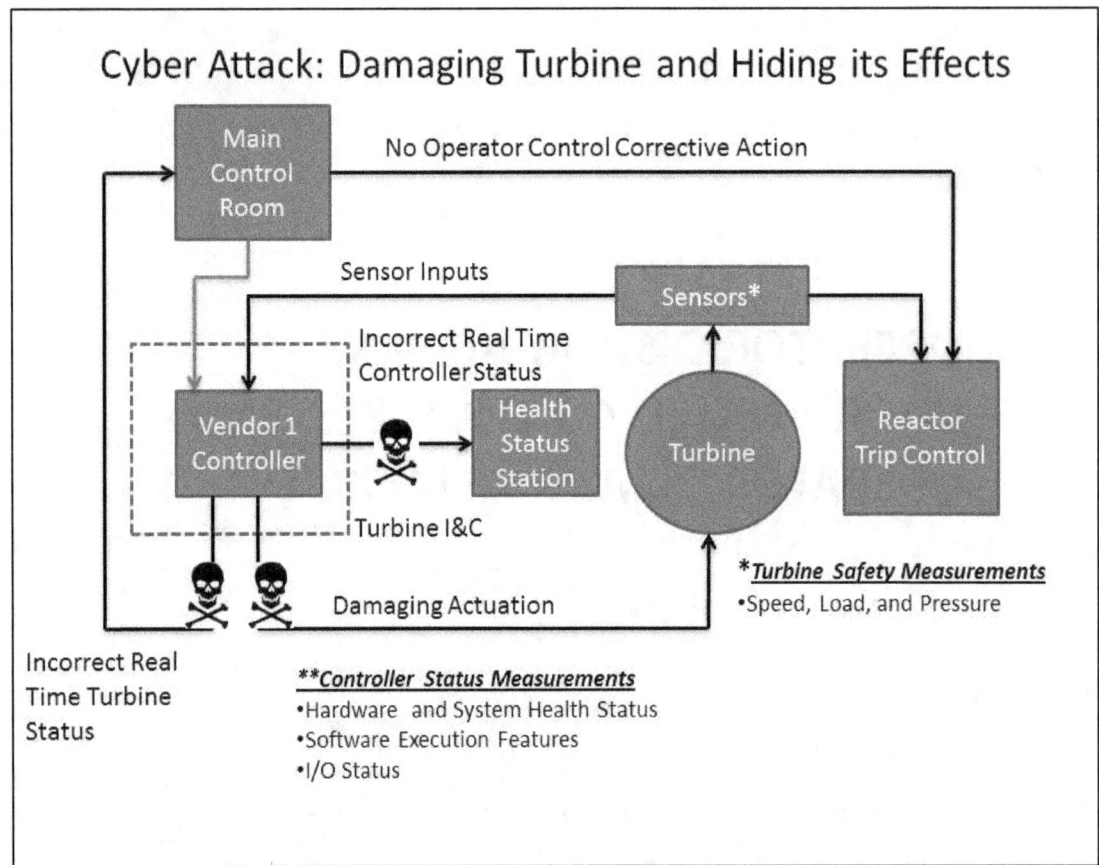

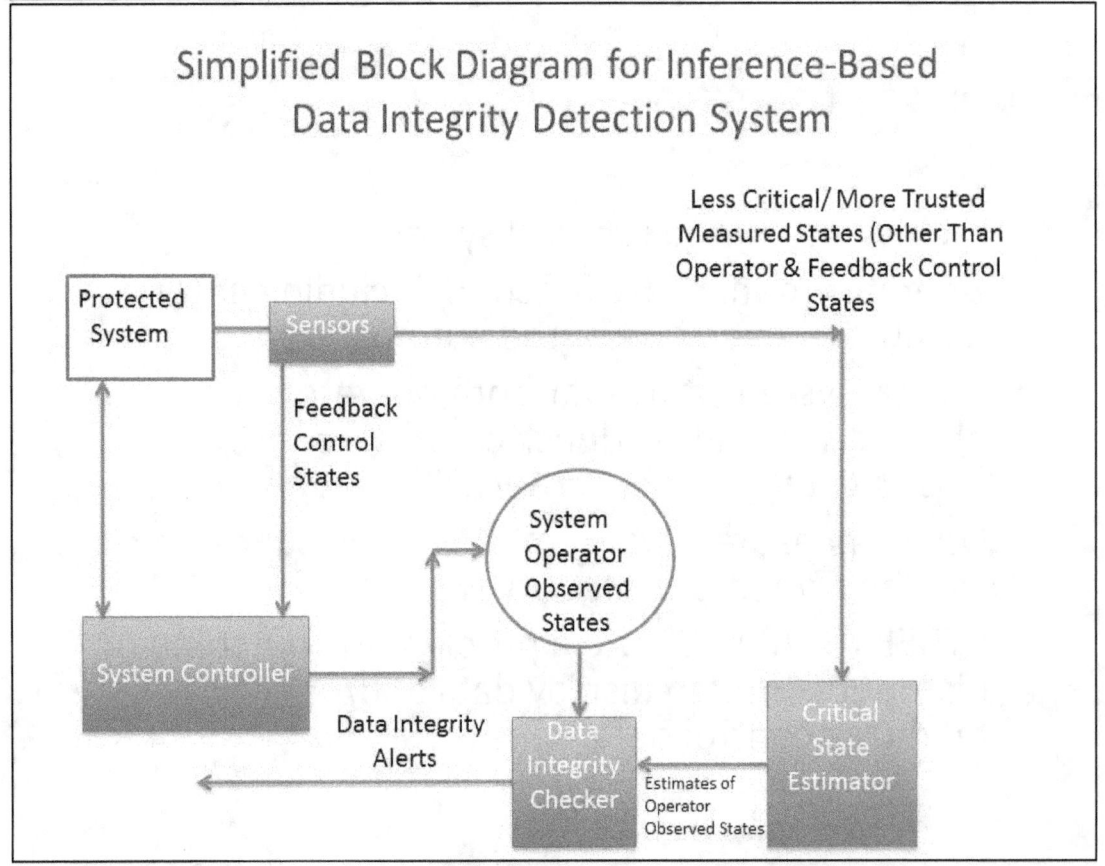

EXAMPLE

Regulating a Linear Physical System (1)

- Linear physical system represented by difference equation
- $\underline{x}(k+1)=\underline{Ax}(k)+B\underline{u}(k)+\underline{\omega}(k)$ where
- $\underline{x}(k)$ is an n vector representing the system state during discrete time interval k
- A is the n x n system state transition matrix
- B is the n x g system control matrix
- $\underline{u}(k)$ is the g vector control signal
- $\underline{\omega}(k)$ is system input noise

Regulating a Linear Physical System (2)

- System measurements are represented by:
- $y(k) = C x(k) + v(k)$
- Where $y(k)$ is a m vector of measurements at time interval k
- C is a mxn measurement matrix
- $v(k)$ is an m vector representing measurement noise

A Simulation Model for Regulating the States of the System

- To facilitate evaluating the data consistency cyber security design pattern:
  - Simulate a linear system controller to sustain the states of a system at designated levels
  - Optimal Regulator Solution (LQG) utilized for simulation
    - White Gaussian noise
    - Separation Theorem
    - Kalman Filter for state estimation
    - Ricatti Equation-based controller for feedback control
  - Controller feed back law based upon variances of input noise, measurement noise and the A,B and C matrices of the system dynamics model

Example State Equations and Noise Assumptions

A = [1,   1. -.02, -.01

  .01, 1, -.01,   0

  .2, .01,   1,   1

  -.01, .02, -.01, 1];

B = [0 , 1 , 0 , 0];

Operator Observed (less trusted):

C = [1, 0, 0, 0];

Related States (unobserved by operator, more trusted):

C2 = [0 1 0 0; 0 0 1 0; 0 0 0 1]

K1 = 0.25;   process noise variances for each of the states

K2 = 0.25;   sensor noise variances for each of the measurements

Simulated System Operation for Regulation of a State Component at 500

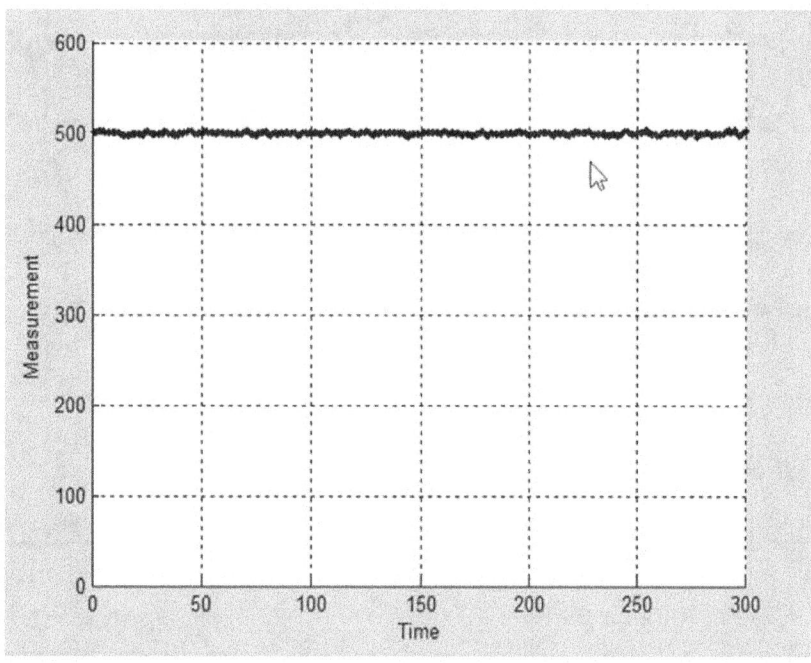

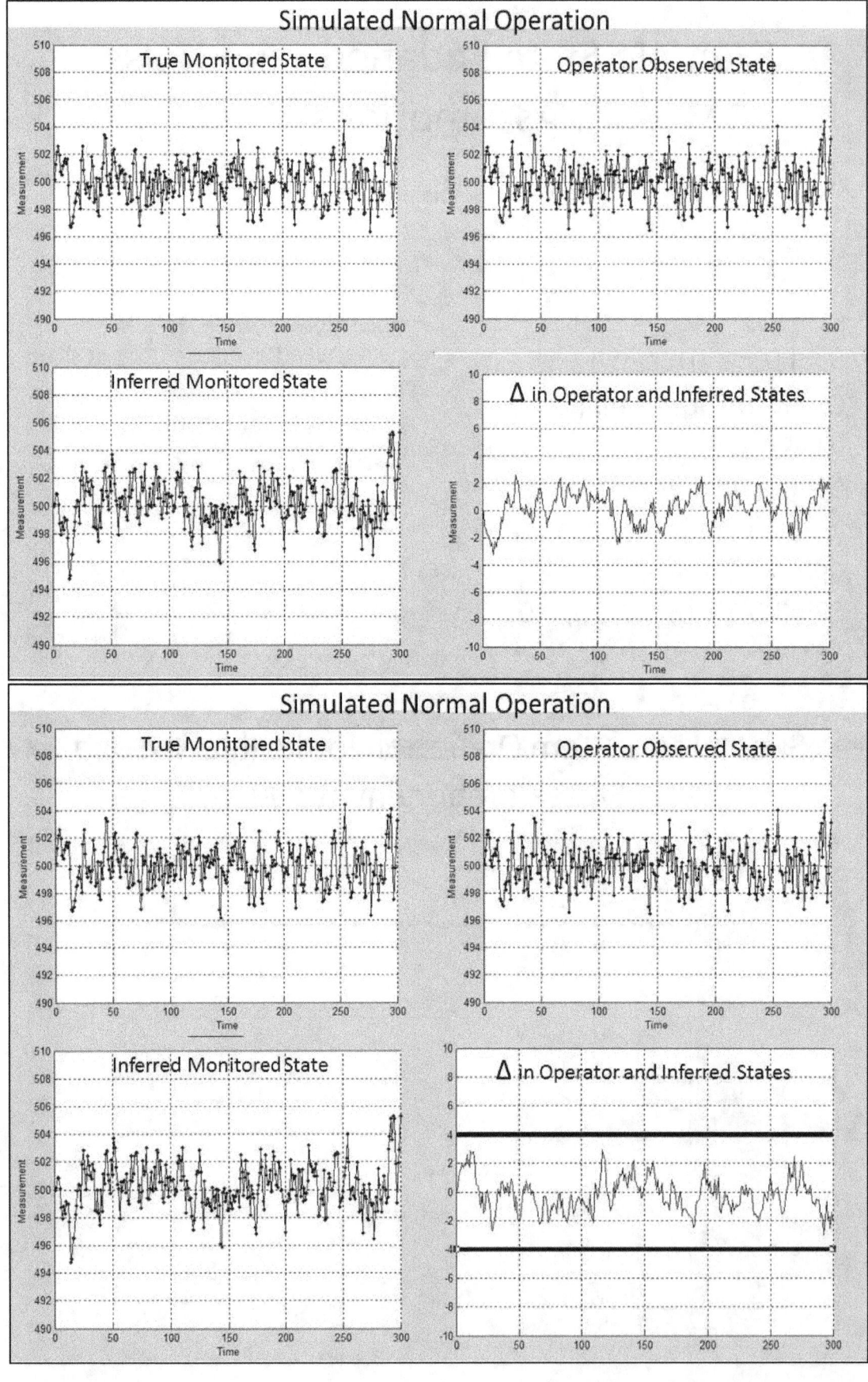

REPLAY ATTACK TO CAUSE ERRONEOUS OPERATOR ACTION

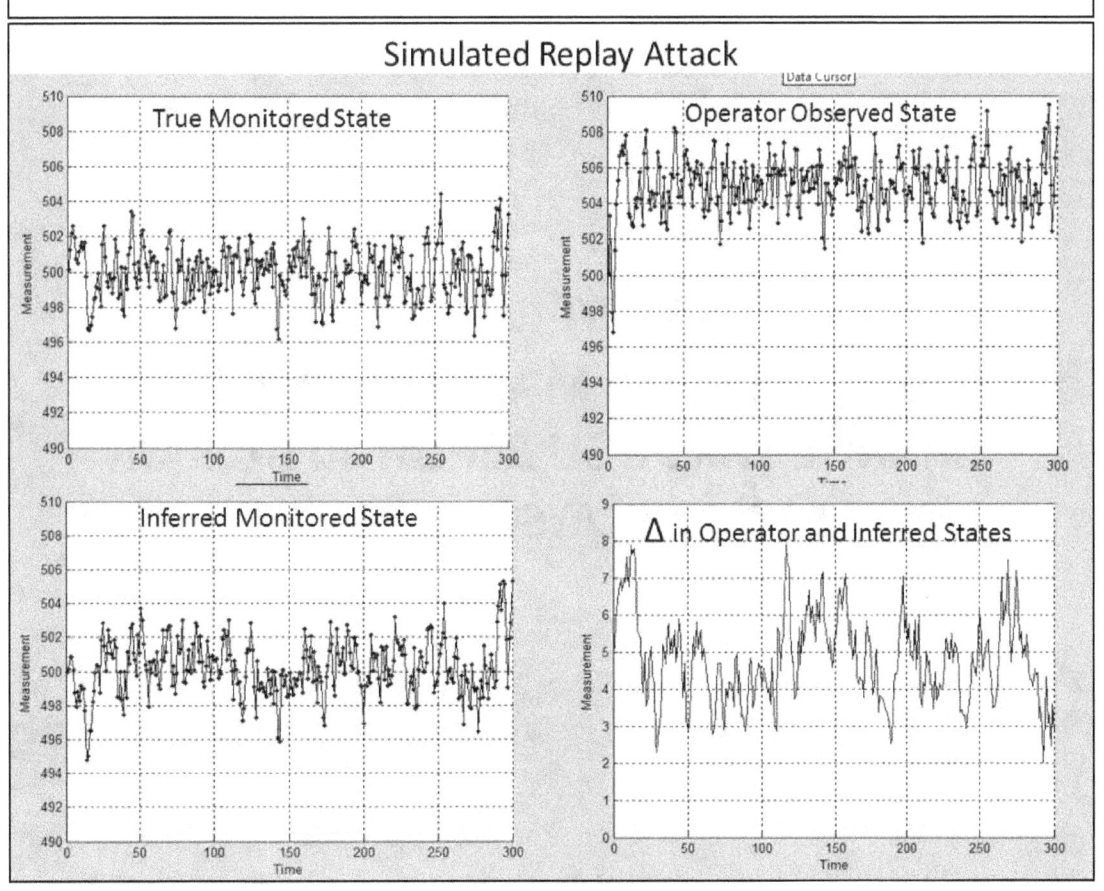

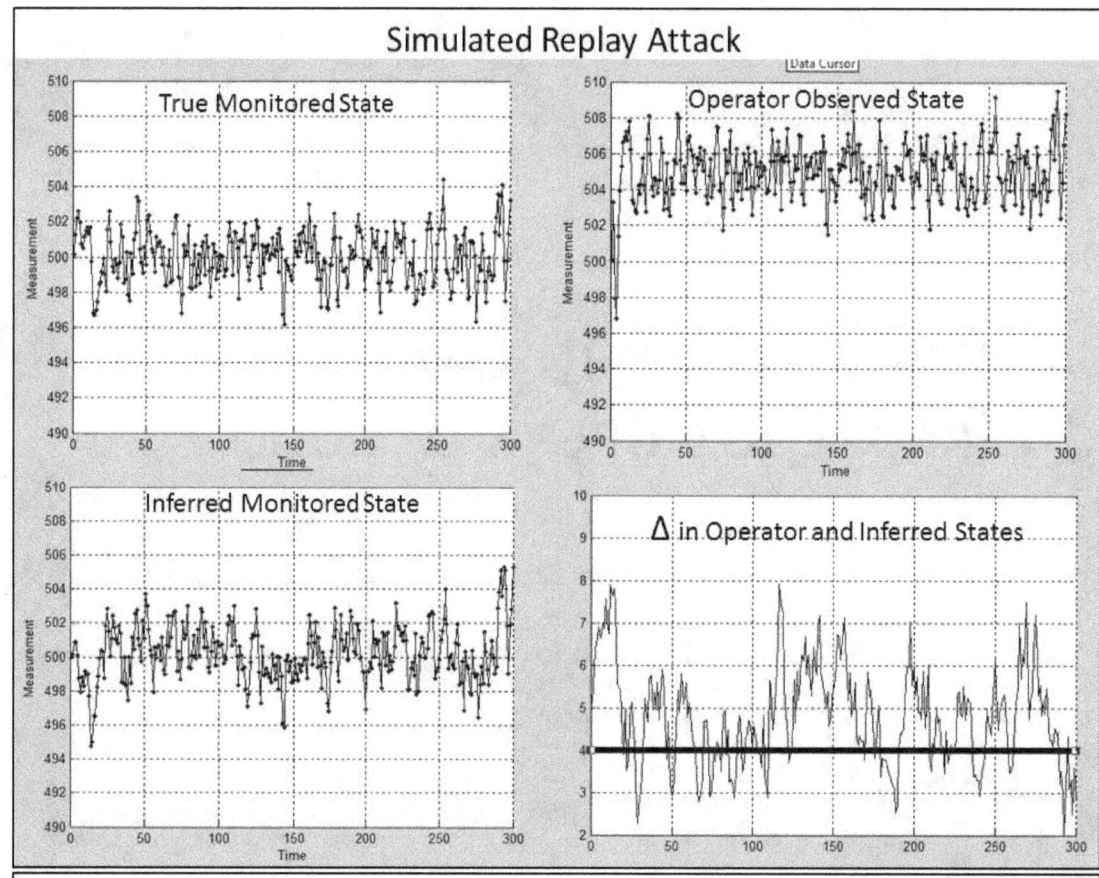

ATTACK TO ADJUST REGULATOR OBJECTIVES AND MASK THE PHYSICAL CHANGE THROUGH REPLAY ATTACK ON OPERATOR DISPLAYS

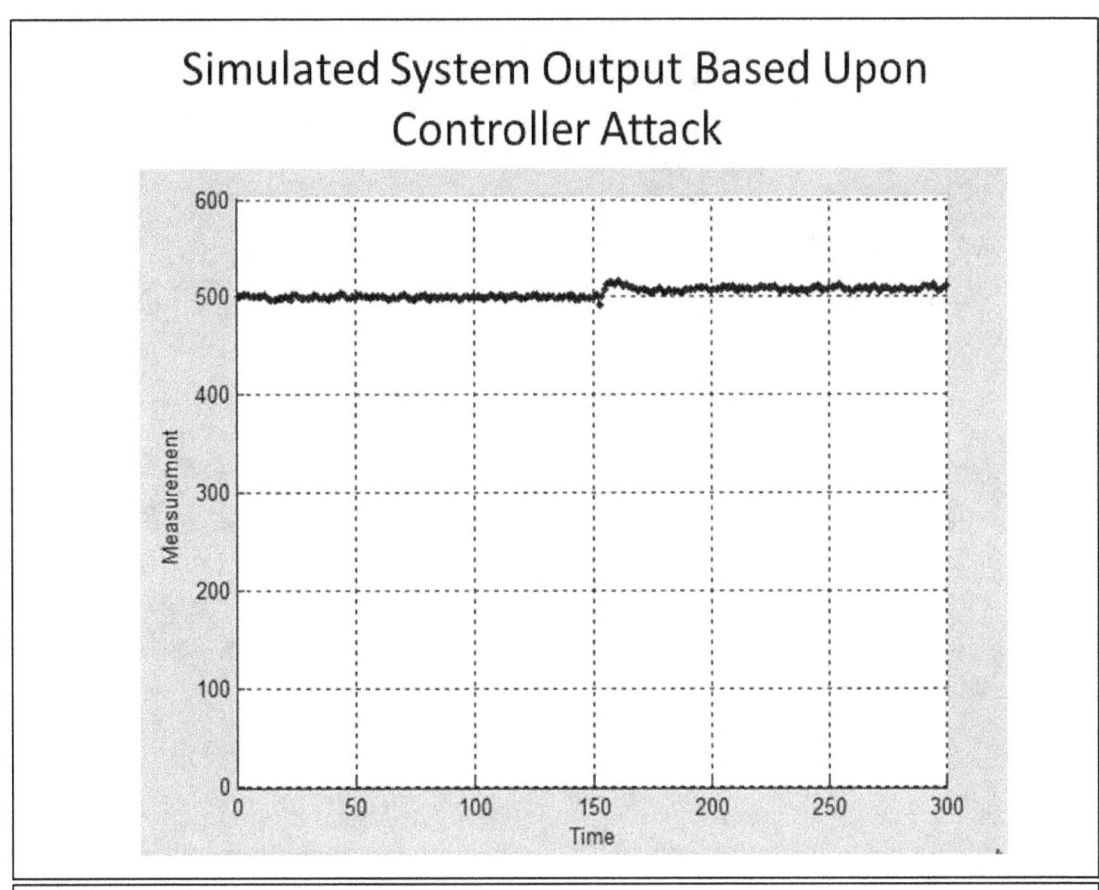

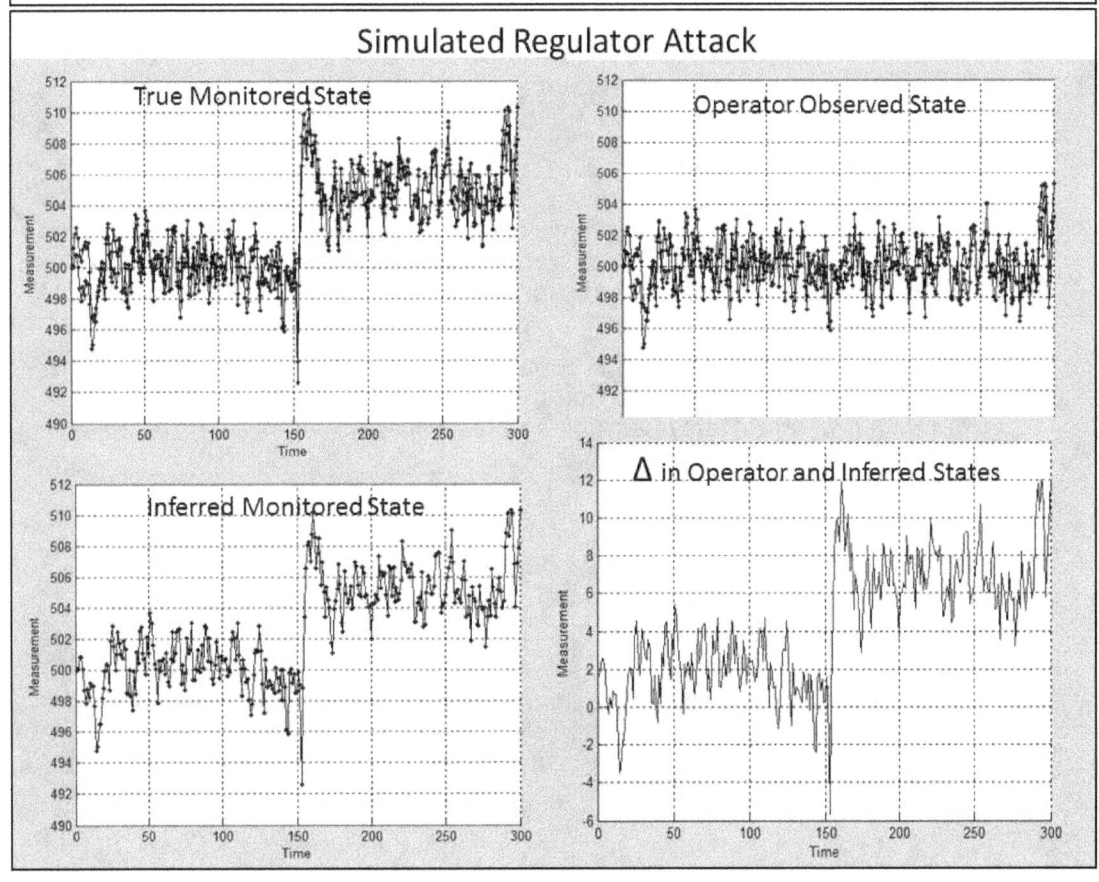

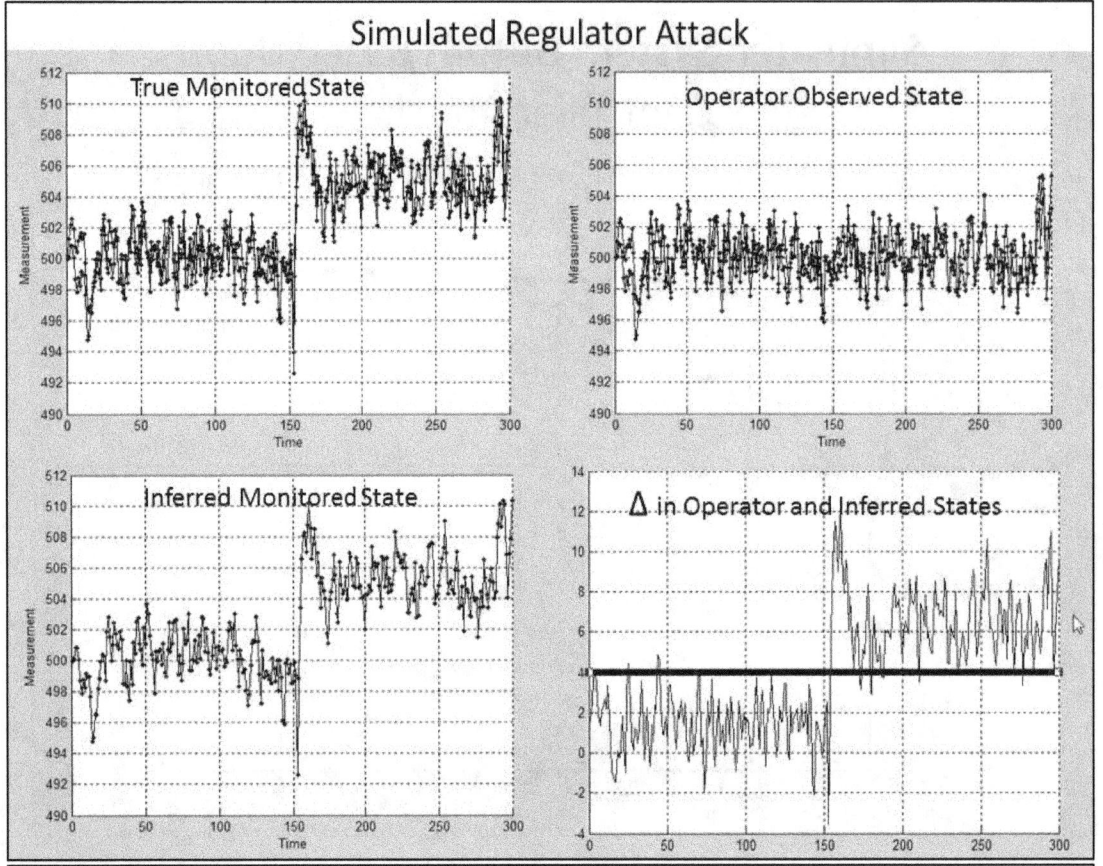

Metrics

- As a practical matter, cyber attack detection/response for mission critical physical systems will need to be tuned to have virtually no model-predicable false alarms for initiating significant responses, such as shut down (for emphasis referred to as "zero" model-based false alarms), while also promising "zero" missed detections.

- Equivalently, sensor accuracy and corresponding detection algorithms must permit use of attack detection thresholds that are greatly distanced from both normal system operation and system operation regions that result in unacceptable consequences

- In order to determine detection thresholds and the corresponding false alarm and missed detection rates, operational data collections would need to be used to build upon model-based analysis, serving to account for shortfalls in system models.

- Detection algorithms and criteria that cause delays in initiating responses must account for how long a system can operate in a region of the state space before an important response is too late

Sliding Window Detection

- For our example, a sliding window detection algorithm is used for integrating over the time series of the "N" most recent individual point detections, each based on a threshold test

  - A cyber attack is declared upon detecting m threshold violations over N detection opportunities
  - Increasing m and N serve to reduce over-reaction to individual estimates resulting in threshold violations, thereby reducing false alarm rate at the expense of potentially increasing the missed detection rate and delaying detections

- More specifically, given a time series of individual point detections, determined by comparing a time series of the most recent state estimates, $x_1, x_2, x_3....x_N$ to an alarm threshold, **th**

- If $x_i >$ **th**, increment g by 1, where:

$$g = \sum_{i=1}^{N} (x_i > \textit{th})$$

- For the example, within a time series consisting of N state estimates each compared to threshold criterion **th**, if g > N/2 a cyber attack is declared.

"Zero" False Alarm Thresholds

150,000 point simulation

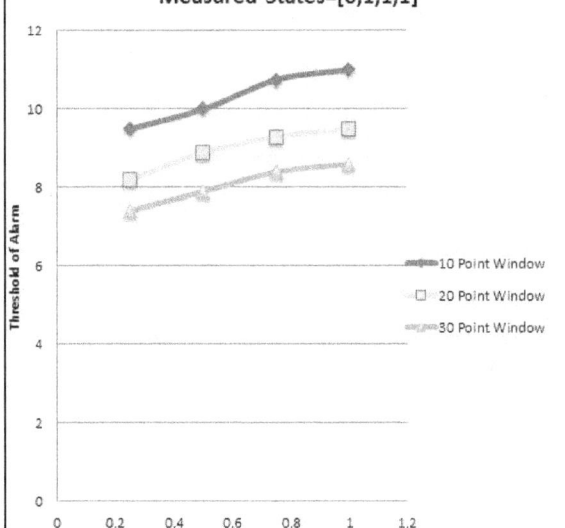

"Zero" False Alarm Decision Threshold; Measured States=[0,1,1,1]

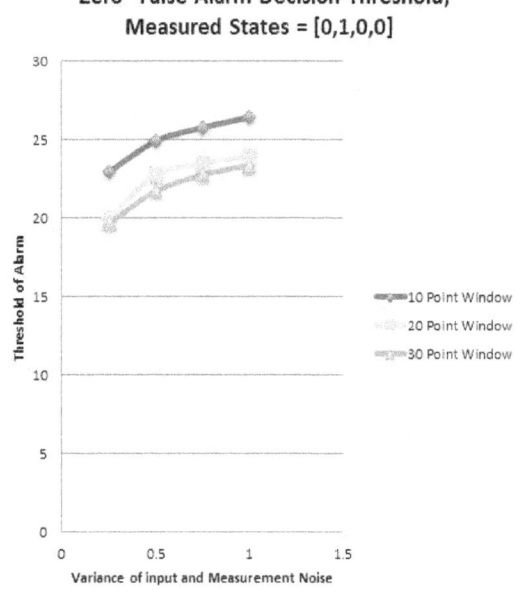

"Zero" False Alarm Decision Threshold; Measured States = [0,1,0,0]

"Zero" False Alarm Thresholds

150,000 point simulation

"Zero" False Alarm Threshold; 10 Point Window / Minimum 10 Second Delay

"Zero" False Alarm Threshold; 30 Point Window/Minimum 30 Second Delay

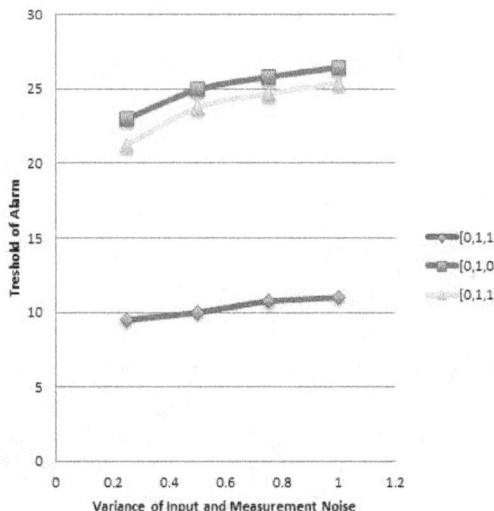

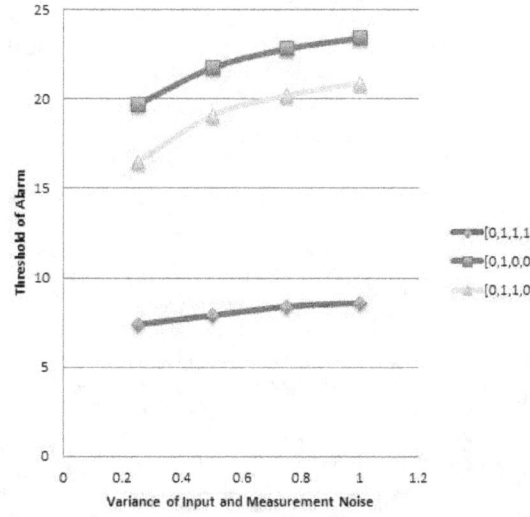

Design Sensitivity Analysis

- Decision Thresholds vs sensor accuracy – ~20-30% change in threshold value over sensor accuracies (variances) ranging from 0.25 – 1
- Decision Thresholds vs selection of states used for inferring critical state(s) values – ~200-300% change in threshold value over state measurement range of [0,1,1,1] to [0,1,0,0]
- Decision Thresholds vs delays in detection (length of sliding window)-10-20% change in threshold value over a 10 – 30 second sliding window detector
- Design range of threshold values comparing the worst case (lowest thresholds) and best case designs (highest thresholds) for achieving "zero" model-based false alarm/missed detection rates – ~400% change from worst accuracy, least states measured, longest sliding window detector to best accuracy, most states measured, shortest sliding window detector

Real World Example: Gas Turbine

- RPM – 3600
- Measurement Error – 1-2 rpm ✔
- Data Interval - 40msec ✔
- Trip Threshold – ~10% rpm deviation ✔
- First estimate of augmenting sensor-based Trip Threshold - ~1% rpm deviation ✔
- Suitable spacing between attack detection thresholds and operating in regions with significant adverse consequences, permitting "zero" model-based false alarms/missed detections ✔
- Multiple triplex sensors – A/D converters and processor interfaces on a single board ✖

Relating Detection Thresholds, System Responses, and Acceptable False Alarm Rates

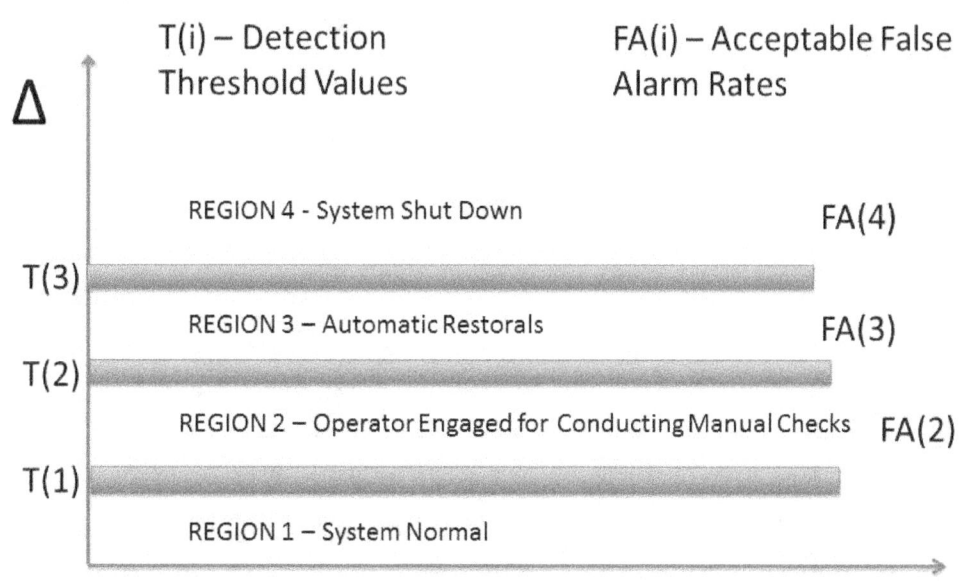

T(i) – Detection Threshold Values

FA(i) – Acceptable False Alarm Rates

Δ

REGION 4 - System Shut Down   FA(4)

T(3)

REGION 3 – Automatic Restorals   FA(3)

T(2)

REGION 2 – Operator Engaged for Conducting Manual Checks   FA(2)

T(1)

REGION 1 – System Normal

ATTACK ON CRITICAL SENSORS' OUTPUTS

Design Pattern Based Upon Cyber Security
Extension of:
T. Kobayashi, D. L. Simon, Application of a Bank
of Kalman Filters for Aircraft Engine Fault
Diagnostics, Turbo Expo 2003, American
Society of Mechanical Engineers and the
International Gas Turbine Institute, June, 2003

Simplified Block Diagram for Sensor Attack Detection System

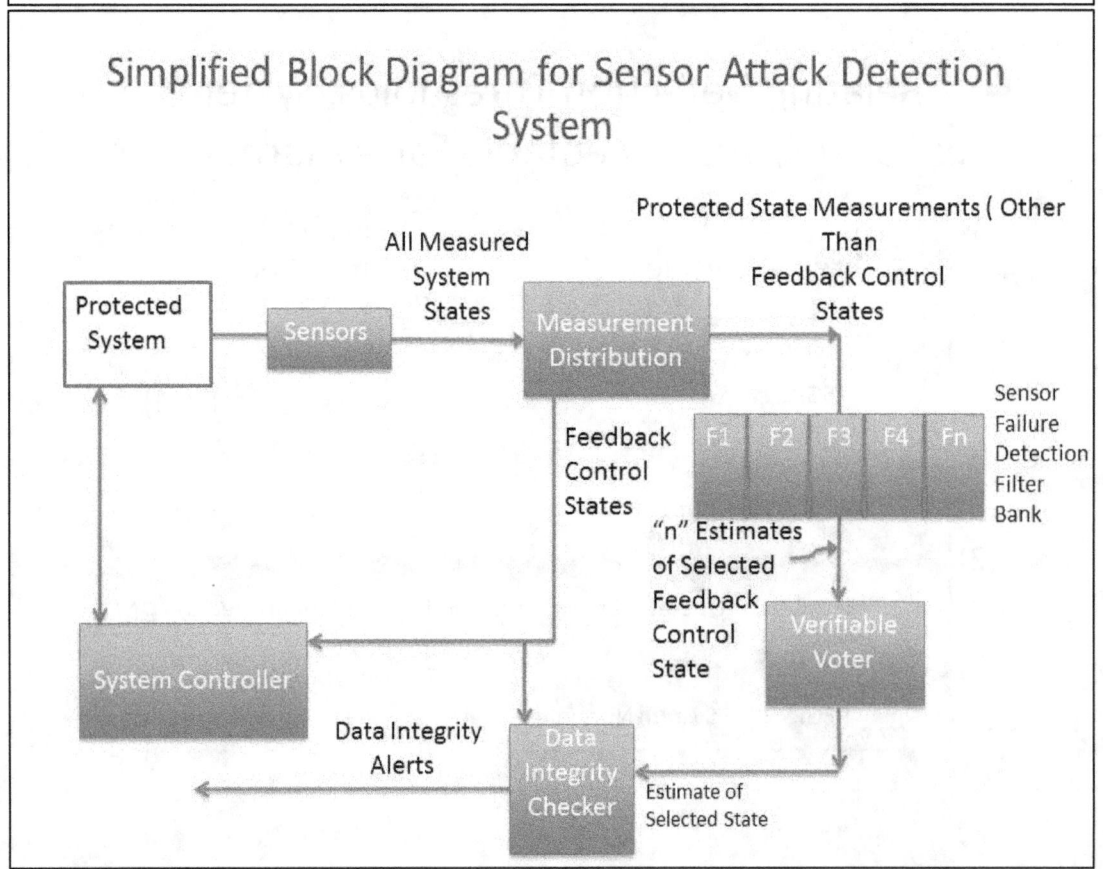

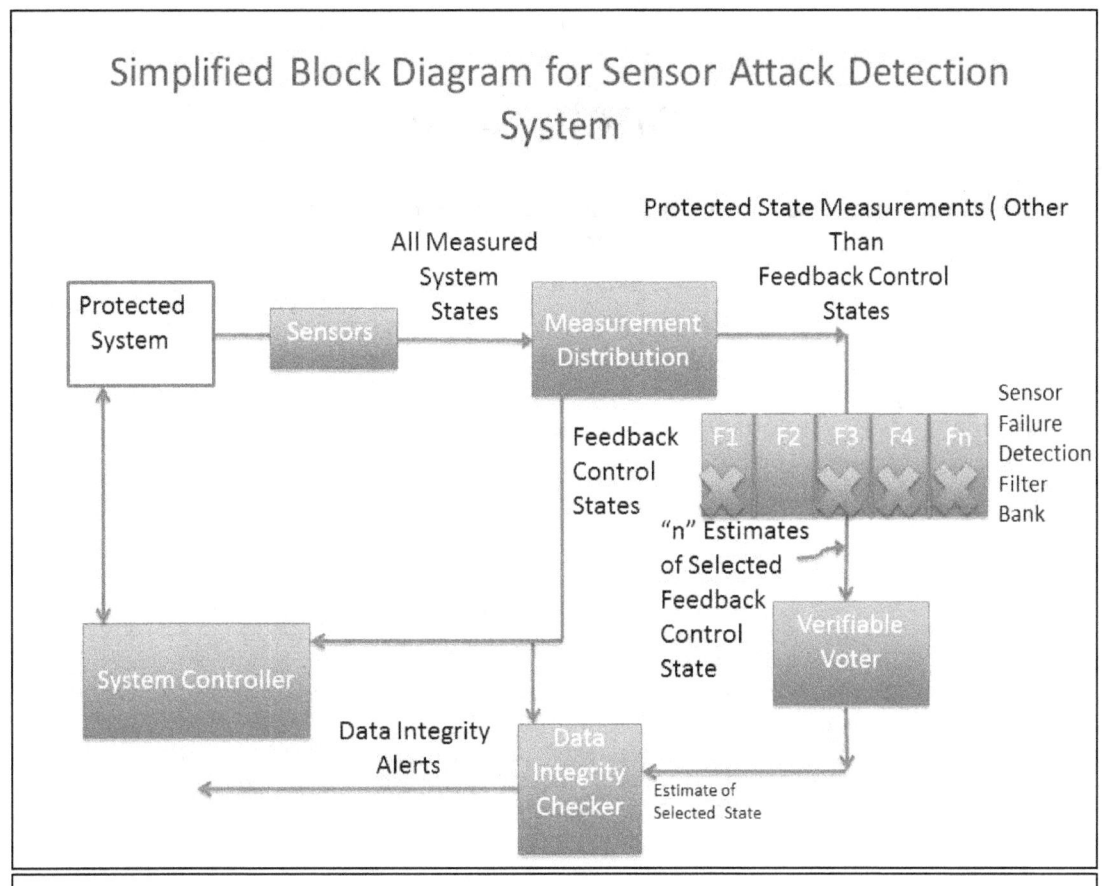

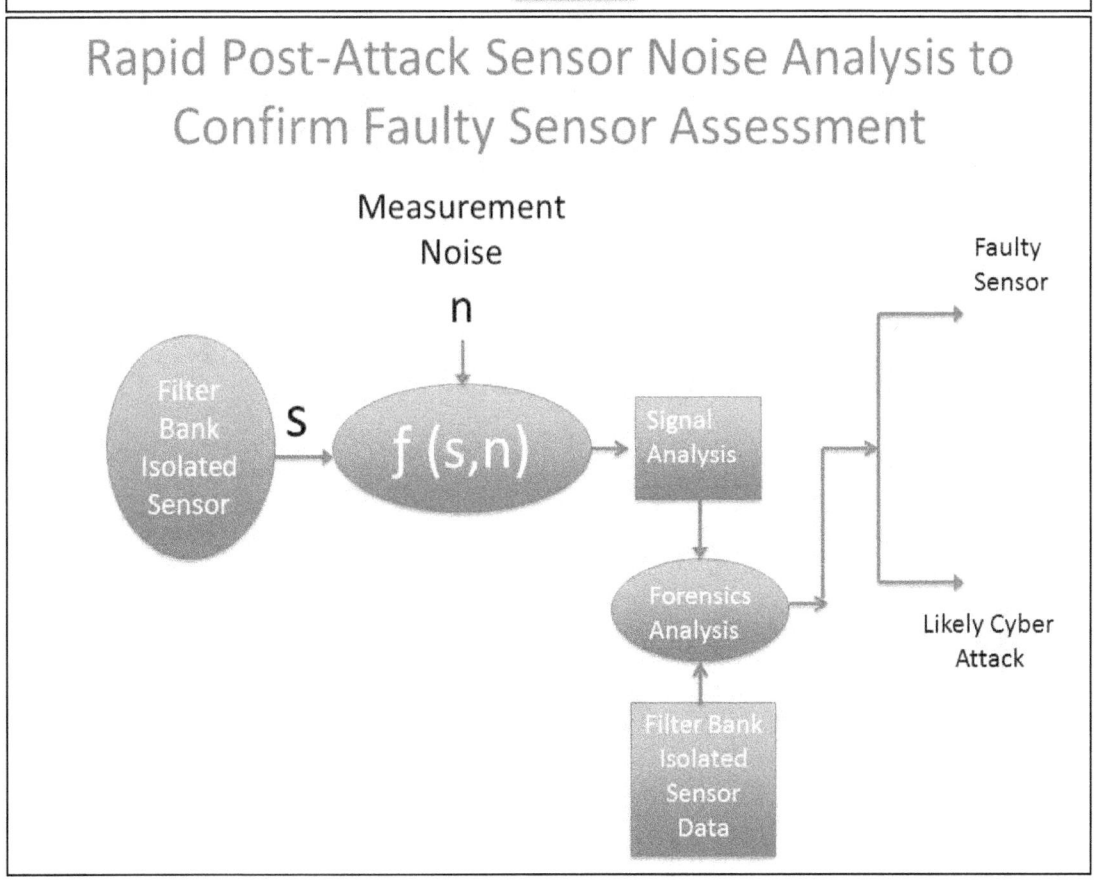

Conclusions

- Data consistency checking design patterns can potentially make an important contribution to cyber security of physical systems

- Past work in fault-tolerant and automatic control systems provides a starting point regarding solutions and knowledge to draw upon, although specific solution designs will need to be implemented in a manner that is sensitive to the issues surrounding cyber attacks

- Development of actual solutions will require system activities in:
  - System dynamics modeling
  - State estimation
  - Security-focused analysis regarding attack scenarios, protection needs, more trusted and less trusted components, and sensors and measurement characterization
  - Distributed security solution designs that serve to complicate, and hopefully deter, attacks
  - In-field data collections regarding selection of detection thresholds and responses to achieve acceptably low false alarm/missed detection rates

9. Challenges of Cybersecurity Research in a Multi-user Cyber-Physical Testbed

Thomas Edgar
Pacific Northwest National Laboratory
Richland, Washington 99354
Email: thomas.edgar@pnnl.gov

Tom Carroll
Pacific Northwest National Laboratory
Richland, Washington 99354
Email: thomas.carroll@pnnl.gov

David Manz
Pacific Northwest National Laboratory
Richland, Washington 99354
Email: david@pnnl.gov

I. INTRODUCTION

Deployed Cyber-Physical Systems (CPSs)[5] are often large, complex, and expensive environments that utilize specialized equipment. The equipment is difficult to configure, deploy, and maintain and requires much expertise to correctly instantiate the components into a connected, functional system [3]. The number of individuals with the necessary skill sets is small, and they are expensive due to the high demand. These combined factors have traditionally limited researchers' access and their ability to conduct studies. A multi-user remotely-accessible testbed[6] significantly lowers the barrier of entry by providing researchers with ready access to CPS environments, without them individually needing to invest in the equipment, resources, and expertise to deploy them. Most importantly, users are freed to focus on research and not ancillary system duties.

A multi-user testbed is a shared resource whose equipment acquisitions benefit all users. CPSs are in essence a "system of systems;" a diverse, broad range of equipment is required to research the many faucets of CPSs. Equipment diversity enables modeling realistic environments in multiple domains. Multi-vendor equipment supports interoperability studies and vulnerability assessments. Finally, equipment diversity assists investigators in generalizing results.

Robust scientific experimentation demands repeatable results [4]. When conducted on a testbed, the description of the *system under test* is the testbed configuration and normally includes the equipment, initial configuration, the relationship between devices, and the communication links. Another researcher can then independently verify the results on the testbed.

A multi-user CPS testbed provides significant benefits to cybersecurity research. However, there are notable challenges to creating such a testbed. These challenges are assessed in this paper. The next section summarizes the challenges. The following section discusses avenues for solving these challenges in the context of the power networking, equipment, and technology (powerNET) testbed. Finally, a conclusion section discusses a path forward for the future.

II. CHALLENGES FOR CYBER-PHYSICAL SECURITY RESEARCH USING TESTBEDS

The unique characteristics of cyber-physical systems and a multi-user experimental testbed result in unique challenges for cybersecurity experimentation. Cyber-physical systems have similar issues to general enterprise cybersecurity experimentation such as data sensitivities, experimental separation, and testbed fidelity but cyber-physical systems have additional unique issues. For example, cyber-physical systems add challenges like system scale, physical process simulation, and diversity design. The cybersecurity challenges that have been encountered during the process of designing and implementing a multi-user experimental cyber-physical testbed will be discussed in this section.

Operational IT systems often have data security requirements that require protection. This encompasses Personally Identifiable Information (PII) and Intellectual Property (IP). Cyber-physical systems can also include these issues, but also add problems such as the proprietary nature of the module or architecture of the system and the operational state of their systems. For example, the state estimation models used by control room operators of the electrical grid as well as the data that provides a status of the system can be proprietary. These models and data could provide competitors or threat actors with system weaknesses that could be leveraged for financial gain or exploitation. Due to the data security requirement, a multi-user experimental cyber-physical testbed has the challenge of providing adequate security mechanisms to ensure that only the appropriate users can access data as well as no data leakage of how an experiment may be architected.

Data is not the only protection challenge that must be addressed in a multi-user experimental testbed. Resources must be protected to ensure that one experiment does not impact the results of another. Multiple experiments could be running on the testbed at any time. The effects of one must not impact the others or at a minimum, quantification of the effects of the testbed on an experiment need to be documented for the other experimenters. This must be a part of every testbed used and is necessary for rigorous experimental design. For example, if one experiment is testing the effects of a DoS attack on a system and another experiment is performing a vulnerability assessment of a product it would be incorrect if the second experimenter believed a loss of connectivity to a device was significant to their actions when in reality it was due to the DoS experiment impacting the shared networking resources [2]. Since cybersecurity experimentation often tests abnormal operational cases it is a challenge to protect experiments from impacting others. Also, it is a challenge to quantify impacts when they do occur. This last step is crucial for all assumptions and qualifications made in a testbed.

Cyber-physical systems run the gamut of scale; from small self-contained systems like automobiles up to highly complex systems-of-systems like electrical grids. Providing the capa¬bility to scale a testbed to meet the needs of a broad range of applications is a challenge. The equipment involved in cyber-physical systems are often expensive to buy and configure. The equipment is often hardened for harsh environmental conditions and requires compliance with many safety and reliability standards. Also, the expertise needed to configure and maintain these systems is highly specialized and expensive to acquire.

On top of the scalability challenge is the heterogeneous nature across and within cyber-physical industries. Systems designed for cyber-physical systems are derived from the requirements of the physical processes for which they are monitoring and/or controlling. Therefore, a system in the manufacturing industry is significantly different than one in the transportation industry. This can include different equip¬ment, network architecture, and operational performance and security requirements. However, this challenge goes deeper, and there can be extensive differences even within industries. For example, due to geography constraints an electric utility in a plains state can look significantly different than one that operates over mountainous terrain.

Another issue that can occur due to scaling of experiments is fidelity of the system. Depending on the experimental design, simulated equipment may not reach the fidelity requirements to evaluate the security characteristics of a device. On the other end, an experiment to evaluate the impact of an event on the electrical grid does not require the fidelity of having the actual equipment for the grid. Ensuring a multi-user experimental testbed has the ability to meet the fidelity needs of a broad range of experimentation is a challenge.

Integration of the physical process into the testbed is a closely related challenge intertwined with fidelity. CPS requires a data substrate which is the physical processes they monitor and control. This substrate interplays with the CPS, providing input and reacting to output. It is often difficult if not impossible to replicate these physical processes in a laboratory environment. Therefore, a simulation capability is necessary to provide the physical aspect of CPS. Creating a simulation capability with high enough fidelity to model the real world is challenging.

III. POWERNET: DRIVING SOLUTIONS FORWARD

The power networking, equipment, and technology (powerNET) testbed [1] is an implementation of a multi-user experimental CPS testbed. In this section, powerNET will be introduced and the envisioned path to solve the challenges defined in the previous section. PowerNET is an effort to build a testbed capability that is multi-user, remotely and dynamically configurable, and user friendly.

In order to provide the necessary data and network separation between users and experiments, powerNET uses a variety of technologies. Each user and project are provided with networked shared directories via NFSv4. To provide authentication and authorization services, Kerberos is utilized. Scripts built into the testbed OS images, on startup, retrieve user and project keys to mount the shares and provide access. Virtual LANs are utilized to provide separation between experiment network traffic. Additionally, overprovisioning of shared resources will alleviate cross experimental impacts.

powerNET provides a unique capability to provide scalability and different levels of fidelity. powerNET combines simulation, virtualization, emulation, and real cyber-physical equipment in one testbed. This combination enables high fidelity small scale experimentation with bare metal equipment. However, it can also scale up to medium scale and slightly less fidelity with virtualization and emulation. Lastly, simulations can be run to enable experimentation at large scales. The combination of all three enable a flexible environment that can change based on the needs of the experimenters.

Similarly, powerNET was designed modularly and for dynamic configuration to enable a broad spectrum of research. CPS includes a diverse selection of industries and equipment. While powerNET currently has a focus on a subset of power transmission and distribution applications, its modular design enables expansion into other applications within the power industry and even into other cyber-physical domains (i.e. oil/ natural gas, water/ waste water, transportation, etc.). And due to the heterogeneous architecture of the industries, powerNET is dynamically configurable so as to enable the modeling of a wide range of realistic architectures.

There are multiple avenues to integrate simulation of physical processes into a multi-user testbed. The simplest but least accurate option is to perform complete simulation of the process and equipment. With a higher fidelity, process simulators can be leveraged to generate data files that represent the instrumentation of the physical world. These data files can then be used to generate digital and analog I/O that can be fed into the CPS equipment. However, this method does not create a reactive experiment. The highest fidelity would be to dynamically integrate physical processes into a testbed. This can be done be via a real-time running simulator that can inject digital and analog I/O while also be able to respond to communication from the CPS equipment. The Real Time Digital Simulator, used in the power industry, is an example of such a capability. All three have their uses and are viable options depending on the experimental setup. During experimental design, researchers must be aware of the level of fidelity offered by testbeds with differing configurations and choose the appropriate setup based on experimental requirements. This needs to be an explicit part of experimental setup and design and not an implicit, or perhaps overlooked afterthought.

IV. CONCLUSION

While a multi-user (CPS) testbed has many benefits, some operational challenges must be addressed. The set of challenges defined in this paper are by no means a complete enumeration. The challenges listed are the most pressing that have been analyzed in the development of the powerNET testbed. Some of the challenges discussed are significant and may require research efforts of their own.

In addition to these challenges, there exists a more fundamental generalization issue or external validity problem for all of cybersecurity science. The field still lacks good protocol to quantify how well the demonstration of a security solution in one context would apply to the broader community. Also, the cyber domain is quickly evolving and cybersecurity science still lacks a method to apply research results into predictive quantification of how a solution will stand up to threat evolution.

The powerNET approaches discussed in this paper provide a good starting point in tackling the challenges listed. However, in most cases they do not provide a complete solution to the challenge. It is necessary that further work is performed to enable the full capabilities that are desirable in a multi-user CPS testbed.

ACKNOWLEDGMENT

This research is funded by US Department of Energy and US Department of Homeland Security under contract DE-AC05-76R1-1830. powerNET is a part of the Pacific Northwest National Laboratory Future Power Grid Initiative GridOptics capabilities.

REFERENCES

[1] TW Edgar, DO Manz AD McKinnion, TW Carroll, BA Akyol, PM Skare, CW Tews, and JC Fuller, *Power Networking, Equipment, and Technology Experimentation: Designing a Testbed for Cyber-Physical Security Research*, In 7th Annual Cyber Security and Information Intelligence Research Workshop Association for Computing Machinery, New York, NY, 2012

[2] Roman Chertov and Sonia Fahmy and Ness B. Shroff, ACM Trans. Model. Comput. Simul , *Fidelity of Network Simulation and Emulation: A Case Study of TCP-Targeted Denial of Service Attacks*, Num.1, Vol. 19, 2008.

[3] E. A. Lee, *Cyber physical systems: Design challenges*, Eletrical Engineering and Computer Sciences, University of California at Berkley, Tech. Rep. UCB/EECS-2008-8, Jan. 23, 2008, accessed on Apr. 22, 2011

[4] E. Eide, *Toward replayable research in networking and systems*, in Proc. of the NSF Workshop on Archiving Experiments to Raise Scientific Standards (Archive 10), 2010.

[5] National Science Foundation, *Cyber-physical systems program solicitation*, 2011, accessed on Apr. 25, 2011.

[6] J. Kurose et al., *Report of NSF Workshop on Network Research Testbeds*, Nov. 2002.

Challenges of Cybersecurity Research in a Multi-User Testbed

THOMAS EDGAR, *THOMAS E. CARROLL*, AND DAVID MANZ

Pacific Northwest National Laboratory
Richland, WA

May 4, 2012   NIST Cybersecurity for Cyber-Physical Systems 2012   1

Cyber-Physical System are Complex

- ▶ CPS are large, complex distributed systems, comprising specialized, utilitarian equipment
- ▶ Heterogeneous equipment manufactured by multiple vendors
- ▶ Systems are expensive, difficult to configure, deploy, and maintain
- ▶ Requires expertise
  - ▪ Demand exceeds supply

How to support research in the CPS domain?

May 4, 2012   NIST Cybersecurity for Cyber-Physical Systems 2012   2

Challenges for CPS Research

- Equipment must available and accessible
- "Real" data needs to be available
- Researchers shouldn't be forced to be operational experts
- Experiment in a "safe environment"
- Support for the scientific method
- Enable open science

Multi-user testbeds enable CPS research

May 4, 2012   NIST Cybersecurity for Cyber-Physical Systems 2012   3

Testbeds Support CPS Research

- A *testbed* is platform for experimentation (NSF 2002)
  - *Proof-of-Concept*: Purpose-built for demonstration
  - *Multi-User*: Shared resource pool

- CPS multi-user testbeds should:
  - Be dynamic, flexible, and remotely configurable
    - Researcher-friendly configuration
    - Libraries of scenarios, templates
  - Support concurrent experiments
  - Have broad and diverse pool of real world equipment
  - Be modular, extensible, and scalable
  - Support the research community and open science

May 4, 2012   NIST Cybersecurity for Cyber-Physical Systems 2012   4

Put the Researcher in Control

- ► World-wide accessibility
- ► Researcher-friendly interfaces to configure and initialize resources
  - ■ GUI are adequate for small scale experiment
  - ■ ...inefficient when experiments comprise hundreds of components
- ► Library of common designs, architecture, and designs
- ► Activation should be on the order of hours, not days
- ► Mechanisms to simulate "normal"
- ► ... and to orchestrate events, processes, etc.
- ► Default instrumentation and visualization
- ► ...and other mechanisms to inform and collect system state

May 4, 2012   NIST Cybersecurity for Cyber-Physical Systems 201   5

Concurrent Experiments Demand Isolation

- ► Goal is to make efficient use of testbed resources
  - ■ Concurrently running experiments
- ► Experiments should be isolated from one another
- ► Depending on constraints, minimize shared resources
  - ■ Separate management from experiment
  - ■ Support infrastructure, services duplicated per experiment
  - ■ CPS equipment reserved for a single experiment
  - ■ Virtual machine monitors per experiment
- ► Some resources must be shared
  - ■ E.g., Network infrastructure
  - ■ CPS devices, cannot separate the management from experiment
- ► Effects of sharing must be documented and quantified
- ► Method to reserve testbed for single researcher
- ► Resources returned to initialize state on experiment termination

May 4, 2012   NIST Cybersecurity for Cyber-Physical Systems 2012   6

Sensitive Data Must Be Protected

- Organizations demand we protect sensitive data
  - Architecture and design are often considered proprietary
  - Data often contains system state information
- If data is released, may harm or embarrass organization

- Testbed must enforce access controls on the data
- Obscure experiment designs
- Anonymize data employing a scientifically valid approach
  - Paul Ohm's law: "data can either be useful or perfectly anonymous but never both."
  - Several examples of anonymous data that were re-attributed

May 4, 2012   NIST Cybersecurity for Cyber-Physical Systems 2012   7

Testbeds Facilitate Reproducibility

- *Reproducibility* is the condition that allows a skeptic to independently verify results
- From a theory/model the researchers define a *system under test*
- Description of the system is the testbed configuration
  - What resources were used
  - Initial configuration
  - Connectivity between devices, characteristics of links
  - Operating system images, device firmware
  - Logs and serial, network traffic capture
  - Parameters for simulated components
- Unfortunately, uncertain what this means for physical processes...
- Provide mechanisms for researchers to share experimental designs, data, and documentation

May 4, 2012   NIST Cybersecurity for Cyber-Physical Systems 2012   8

Testbeds Can Enhance External Validity

Pacific Northwest
NATIONAL LABORATORY
Proudly Operated by Battelle Since 1965

- ▶ While scenarios/template greatly enhance external validity
- ▶ ...fidelity, equipment, and scale are challenges
- ▶ Real always best, but not always possible
- ▶ Put the researcher in control of fidelity
- ▶ Combine the real with emulated and simulated
  - Procure broad and diverse set of equipment
  - Federate with other testbeds to gain access to additional resources
  - Emulate and simulate other components
- ▶ Simulation should be scientifically valid and researchers aware of shortcomings
- ▶ Some progress on simulated physical processes
- ▶ Bring everything together for experimenting on large-scale systems

May 4, 2012 · NIST Cybersecurity for Cyber-Physical Systems 2012 · 9

powerNET Features

Pacific Northwest
NATIONAL LABORATORY
Proudly Operated by Battelle Since 1965

- ▶ Project-/program- based access controls

- ▶ Remote configuration/execution of experiments
  - Web application
  - Configure using GUI/declarative language

- ▶ Network emulation/simulation
  - DS, SONET, dial-up, wireless

- ▶ Phasors
  - 9 PMUs from multi-vendor/
    1 PMU development platform
  - 1 Hardware PDC/Many software PDCs possible

10

powerNET Features (cont)

- More than 250 virtual nodes possible

- Energy management system (in progress)

- Advanced metering infrastructure (in progress)

- Compute Cluster
    - 3 nodes with SSDs and Infiniband interconnects
    - Scale experiments to thousands of nodes

- 64 TB high-speed shared storage

Use Cases

- Validation and verification
- Technology assessment and prototyping
- Simulation and modeling
- Training and education
- Demonstration

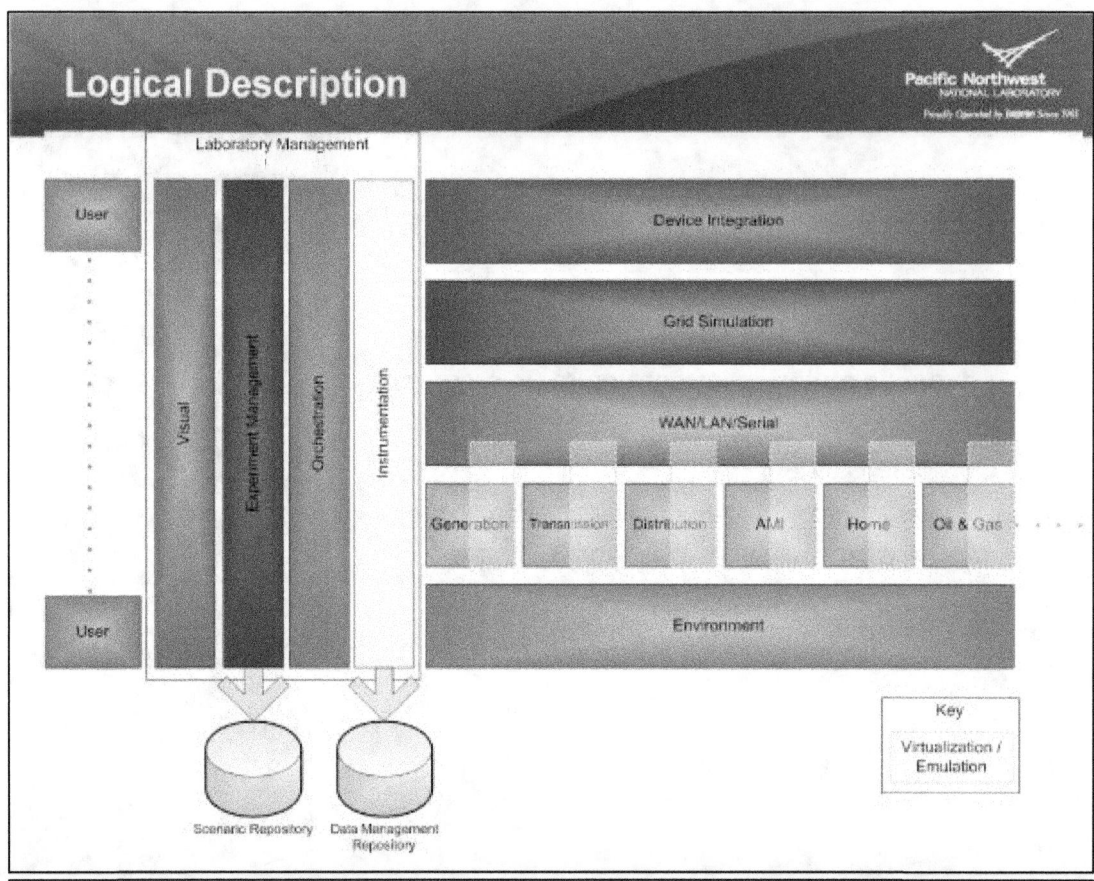

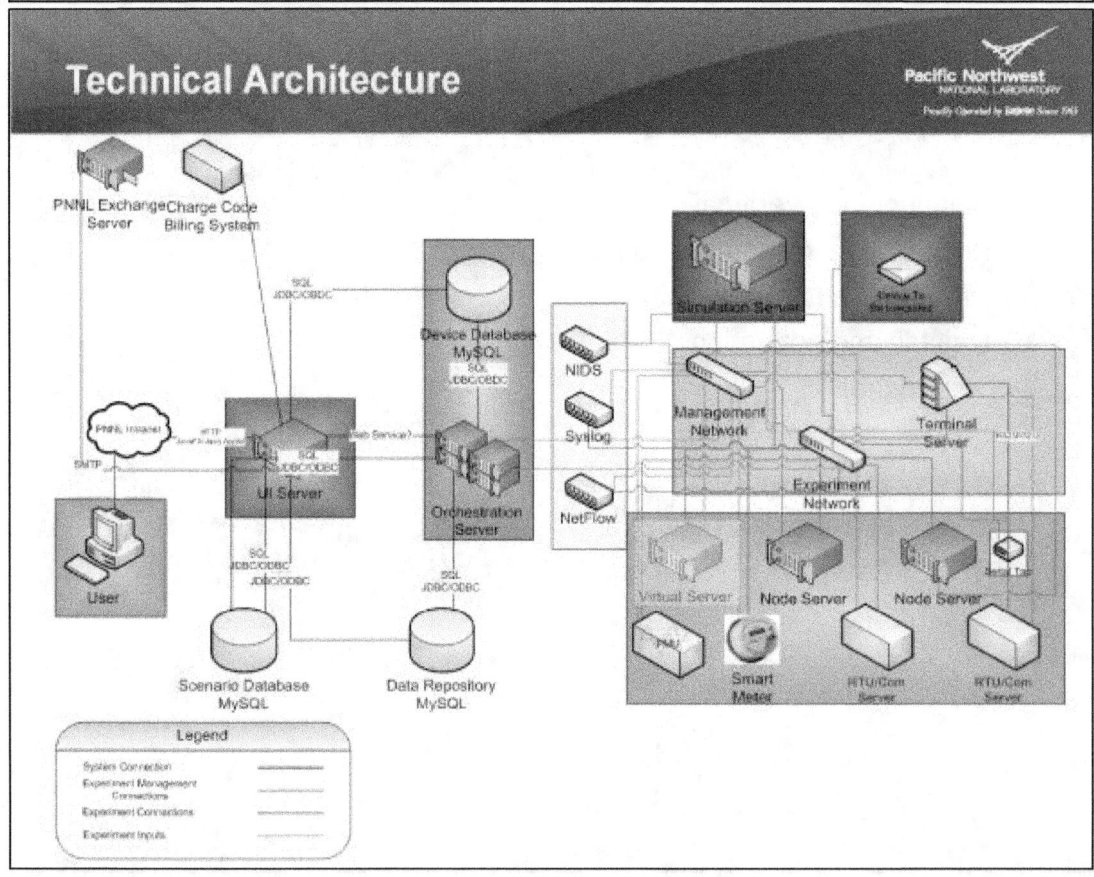

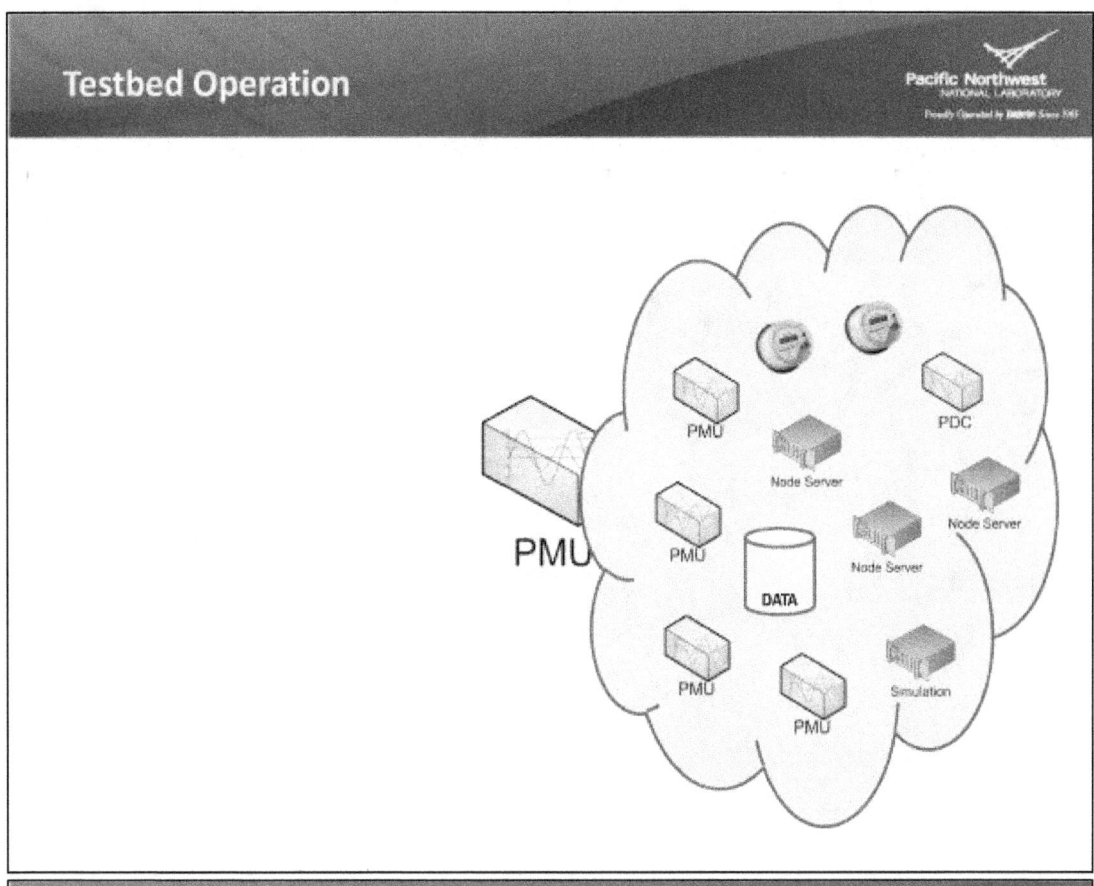

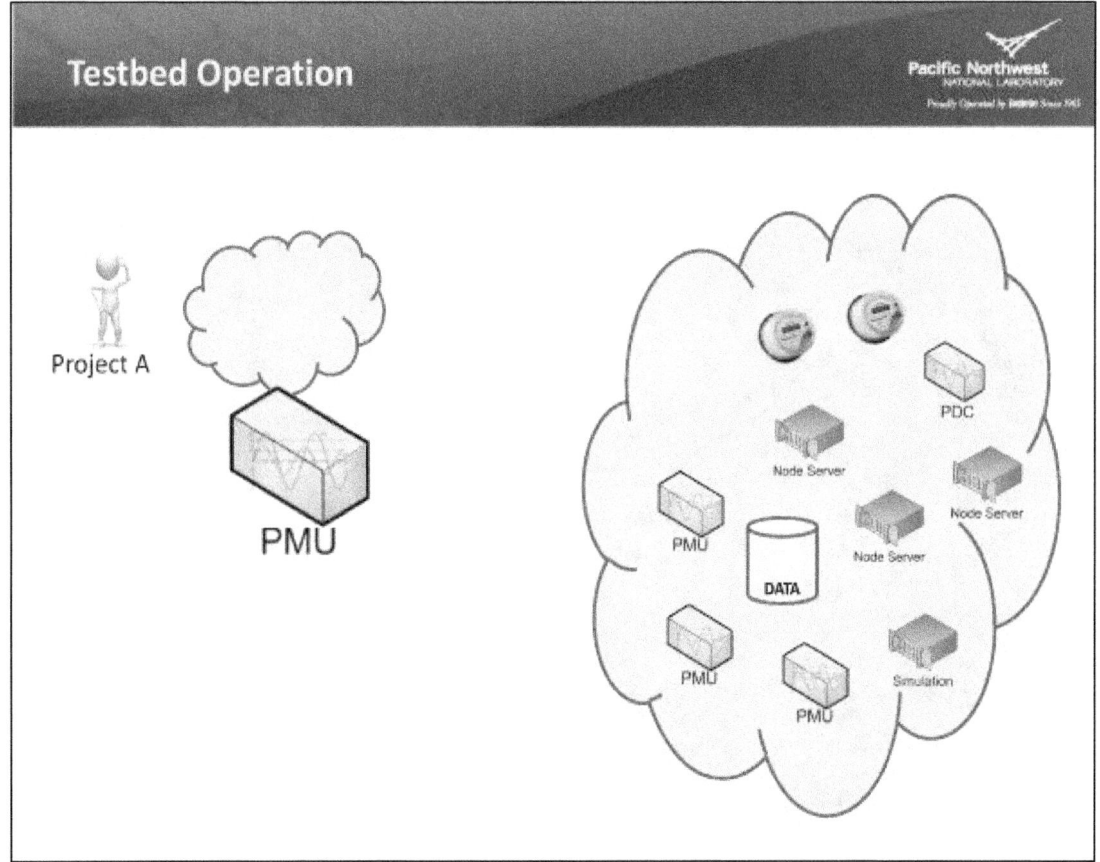

NISTIR 7916

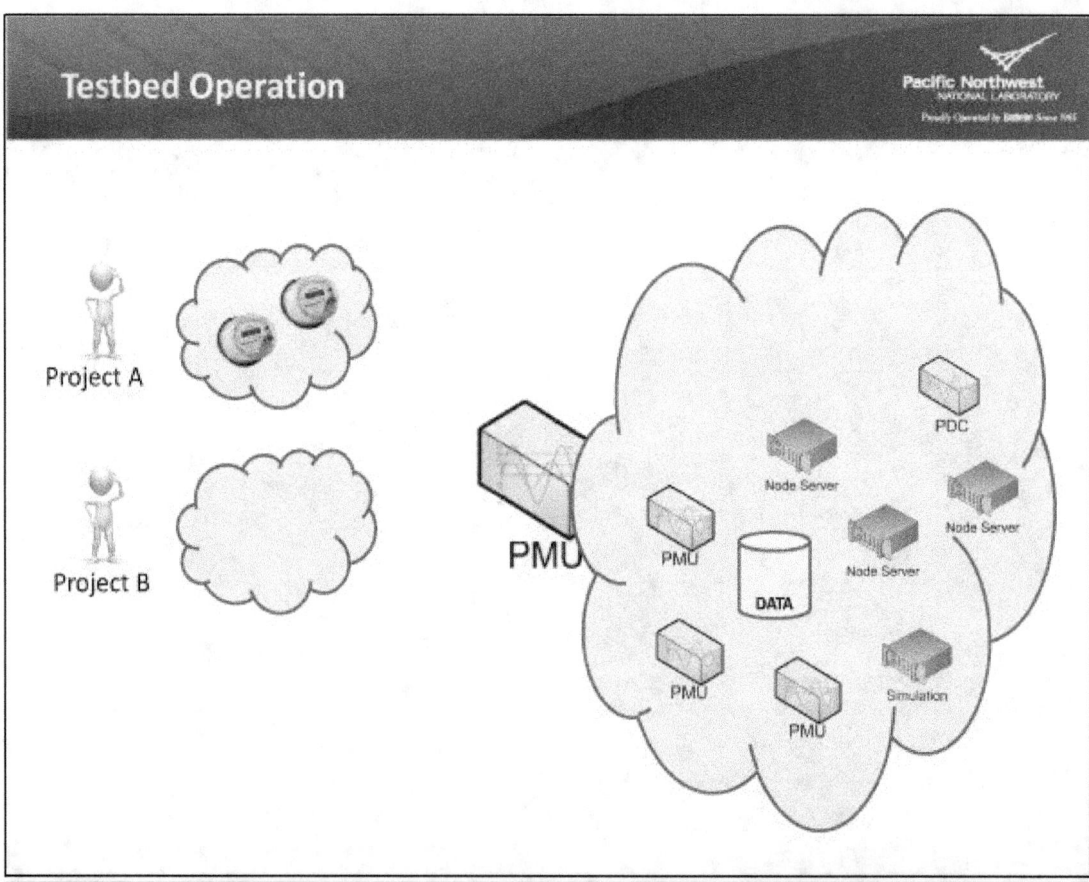

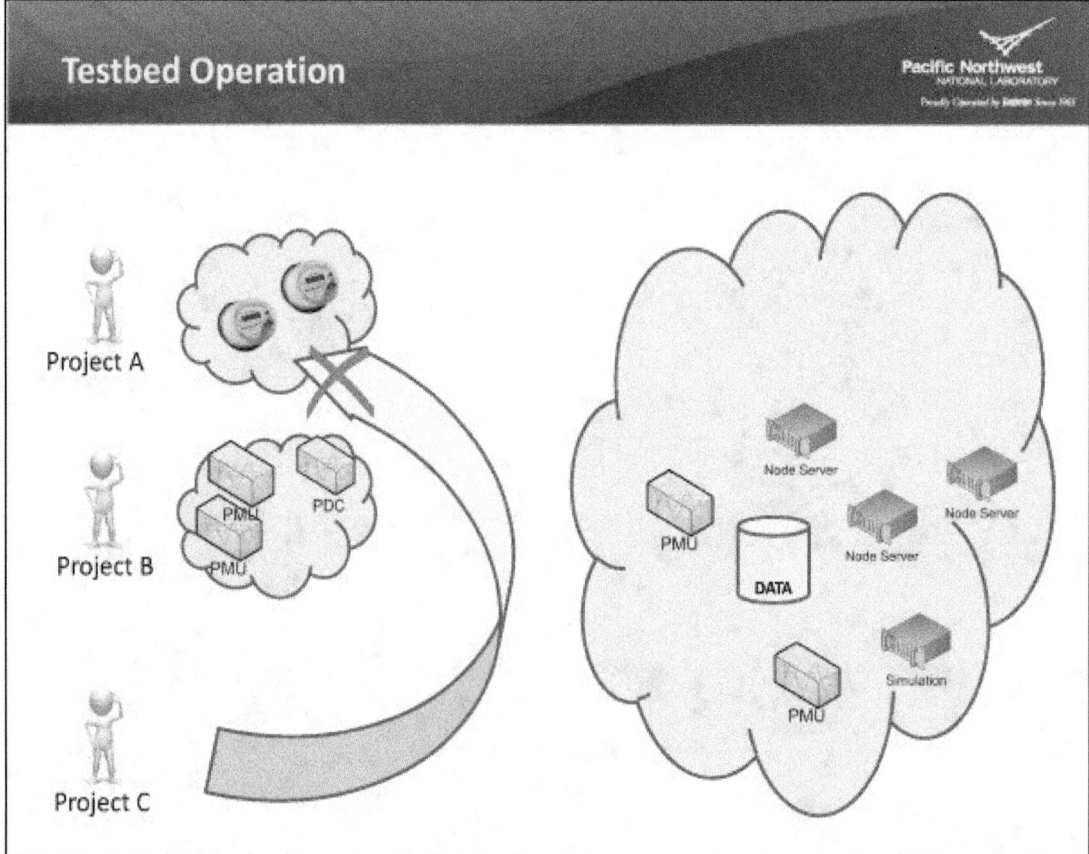

130

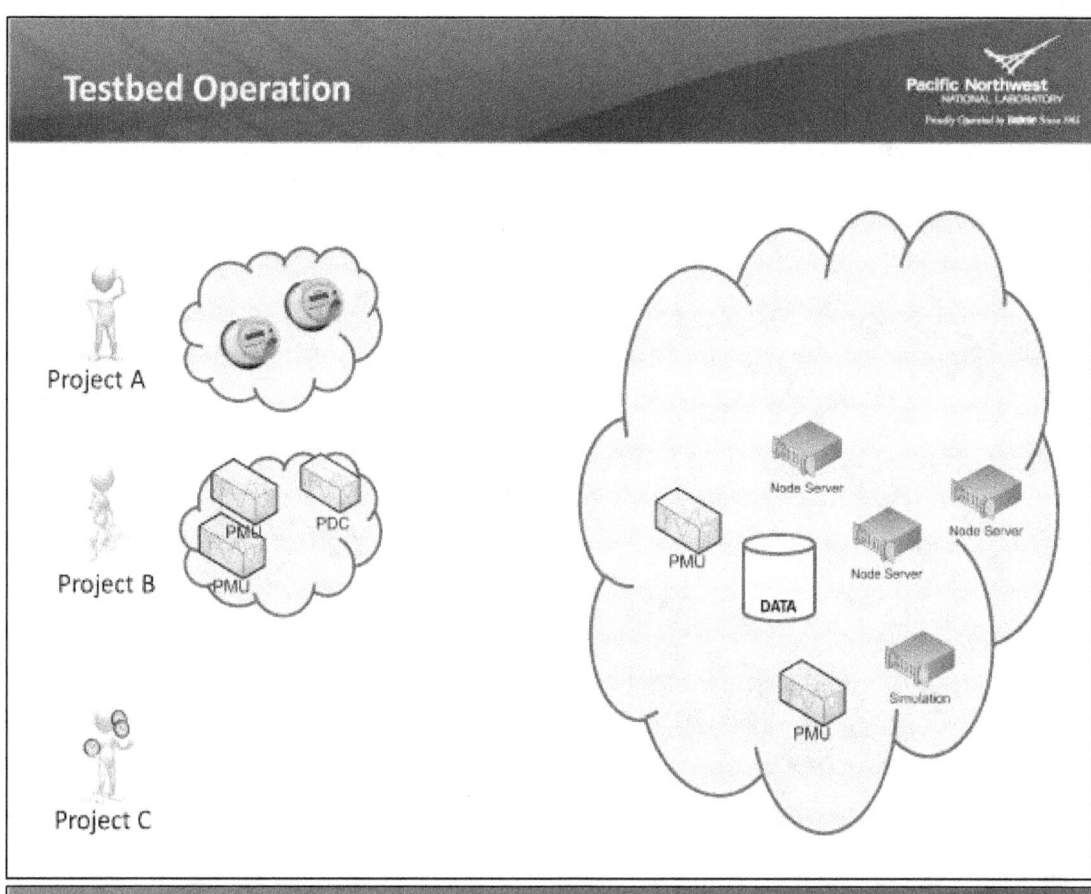

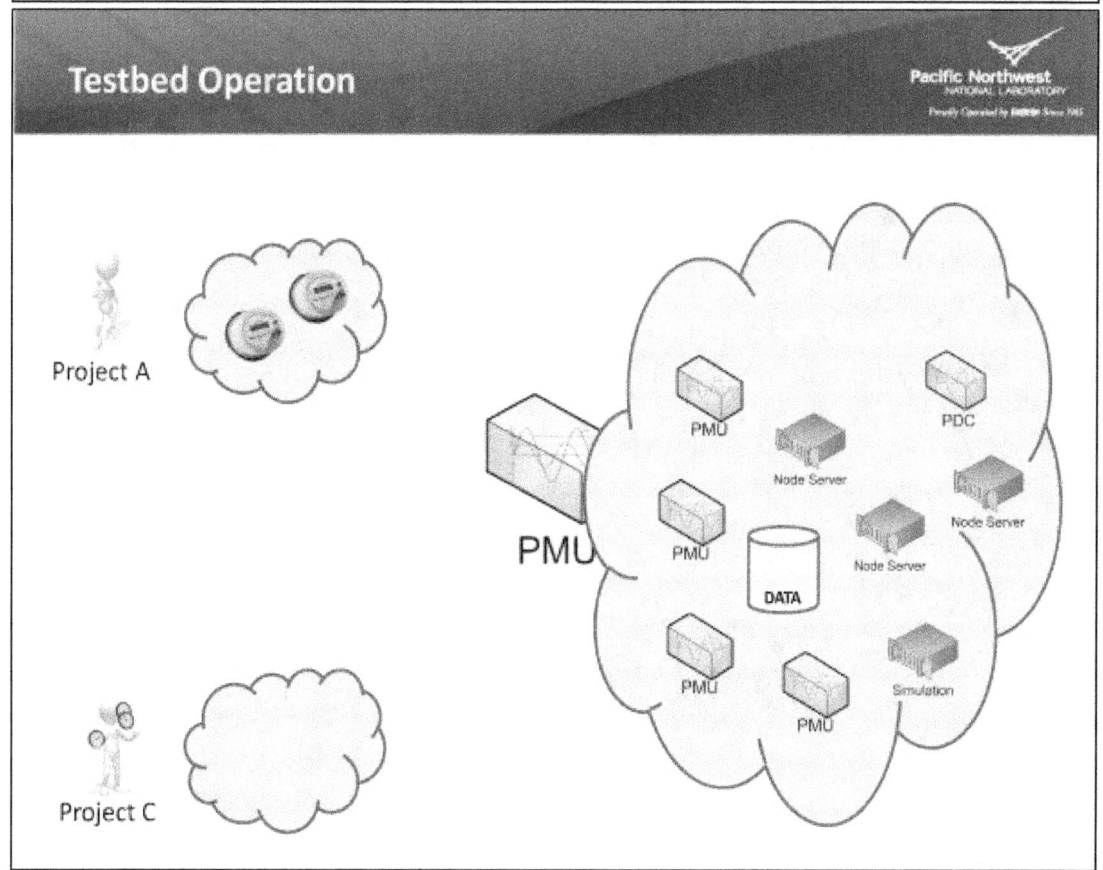

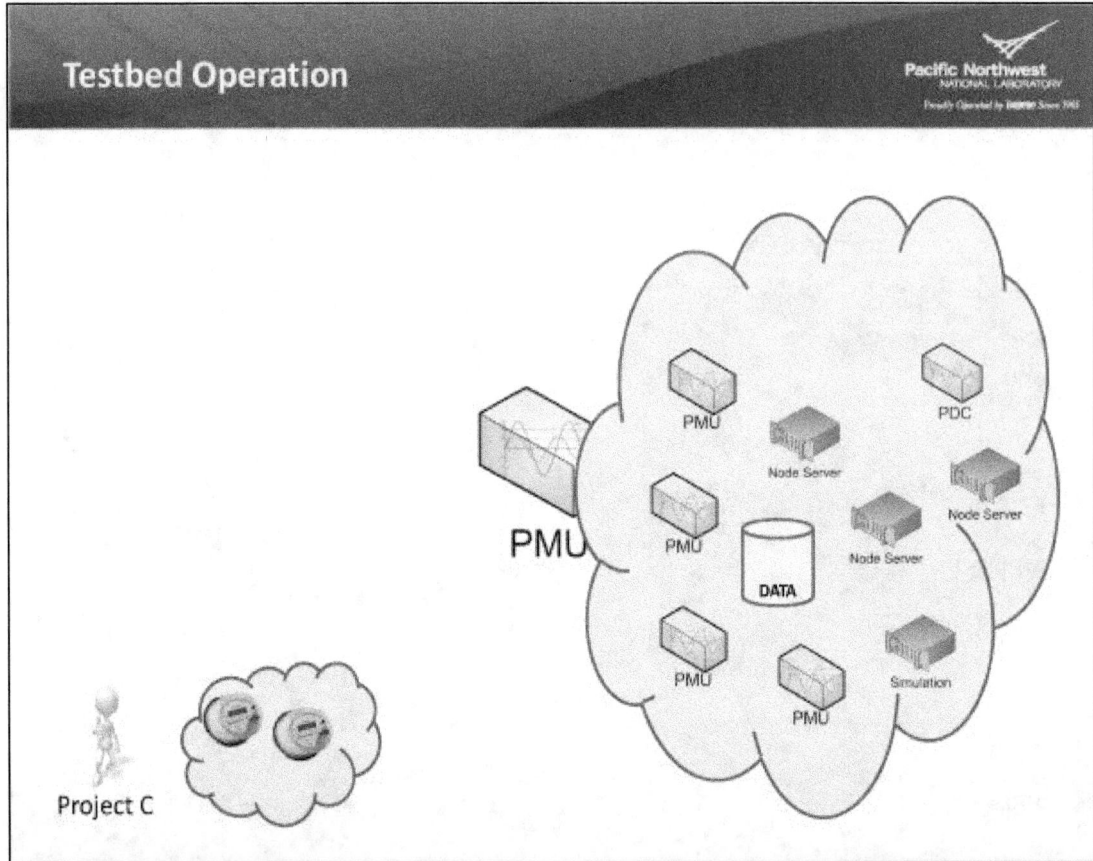

Testbed Operation

PMU

Project C

---

powerNET: Researcher Driven Control

▶ Researchers remote connect to facility
▶ Can configure through a GUI or a descriptive language
  ■ Initial device configuration and impediments
  ■ Provide templates and scenarios
▶ A subsystem for event orchestration

powerNET: Isolation

- Program/Project access controls
- Resources are reserved by and dedicated to experiments
  - ...including virtual machine monitors
- Resources are wiped and re-initialized to a known good state between uses
- Separate control/management traffic from experiment traffic
  - Leverage multiple NICs in devices
- Experiments are isolated from one another using VLANs
- Authentication/authorization resources are duplicated
- Devices cannot communicate directly with one another on the control network
- Data access controls mapped to NFSv4 acls and data confidentiality/integrity provided by NFSv4 and CIFS

Isolation: Reproducibility

- Researchers are free to export their data from the facility
  - Try hard to store data in standard formats
  - Sometimes restrictions on images/firmware
- Community portal/wiki that assists communication between researchers
- Provide archive in support of open science
  - Storage experimental designs, configurations, and data

powerNET: External Validity

▶ A current focus on PMUs, PDCs
▶ …in talks with other equipment vendors
▶ We are federated with other testbeds within PNNL
▶ …and are in the process of federating with DETER and UIUC

Conclusion

▶ Cybersecurity research in CPS has high barrier of entry
▶ Testbeds can ease the burden by providing access and enhancing reproducibility and external validity
▶ Testbeds create new challenges such as isolation and data protection

▶ As a community, we need
   ▪ Scientifically valid approaches for simulating devices and physical processes, synthesizing normal activity and data
   ▪ Access to real data
   ▪ …Scientifically valid approach, with acceptable risk, for anonymizing data

10. Security of power grids: a European perspective

Corrado Leita, Marc Dacier

Symantec Research Labs

Email: {corrado leita,marc dacier}@symantec.com

Industrial control systems (ICS) are rapidly becoming a new major target of cyber-criminals. This was pointed out in multiple occasions by security experts [1], [2] and was confirmed by a recent survey carried out by Symantec [3]. In this survey, 53% of the 1580 critical infrastructure companies that were interviewed admitted to having been targeted by cyber attacks. On average, the surveyed companies admitted to having been attacked 10 times in the last 5 years, with each of these attacks having an average cost of 850k USD. The survey provides a basis for a quantitative estimate of the extent of the problem and implies that the incidents reported by the press over the last several years are nothing but the tip of a considerably larger problem: the vast majority of incidents have never been disclosed. Still, the details of the publicly disclosed incidents give us a better understanding of the underlying issues we face. For instance, a recently discovered malware variant called Stuxnet which has been analyzed at length by Symantec [4] was shown to be part of a highly sophisticated targeted attack aiming at tampering with devices involved in the control of high speed engines, and compromise the associated industrial process [5]. The infection was only uncovered accidentally when an operational anomaly was discovered — Stuxnet has probably been operating undetected since June of 2009 [6]. Stuxnet, and other related threats discovered recently [7], show that industrial control systems are evolving, bringing powerful capabilities into the critical infrastructure environment along with new and yet undiscovered threats.

The power grid infrastructure is a clear example of this evolution. As in other critical infrastructure environments, the idea of interconnecting industrial control systems with other networked computing systems came up only in the last decade, beginning as a method for lowering costs while increasing system efficiency [8]. This convergence is now moving beyond industrial control systems, and the Smart Grid is now being promoted globally as a way to solve problems with energy production, distribution and consumption, to enable energy independence and to combat climate change. Smart Meters, or more generally, the Advanced Metering Infrastructure (AMI), have been aggressively adopted by many European countries. For example ENEL, the Italian utility company, has already deployed over 30 million meters. Similar trends are being followed by other European countries such as France and Netherlands, in which a pilot deployment of 250,000 units will be enriched in the future years to cover 80% of the national installations.

1. Power grid infrastructure and IT security

The convergence between ICS environments and standard IT practices and technologies has important security implications [9], implications which have only been marginally explored by security researchers.

On the one hand, the increasing use of COTS (commercial, off-the-shelf) operating systems (Windows, Linux, etc...) has exposed these environments to attacks, incidents and intrusion techniques characteristic of traditional IT environments.

On the other hand, the employment of standard IT technologies can be seen as an opportunity to access the extensive array of standard IT security techniques (intrusion detection systems, file scanning, standard hardening techniques) and to apply it on these networks. Security techniques honed over many years of practical application can now be used to bear on security issues new to the critical infrastructures.

We claim the trade-off between benefits and associated challenges to be currently imalanced: standard IT security technologies, however robust, cannot protect critical infrastructure as effectively as it is possible in standard enterprise IT environments, given the greater amount of variation in existing control systems and communication protocols as well as the prevalence of older technologies in operation concurrently with newer systems. Moreover, no concrete solutions have been proposed so far for addressing the security concerns associated with new technologies such as AMI infrastructures, where the introduction of basic security primitives (e.g. encryption and authentication in network communications) does not tackle the serious concerns associated to their large scale deployment [10], [11], [12].

1.1. Specific challenges

In order to understand the reasons at the root of the ineffectiveness of standard IT security techniques, we need to look more in detail at the characteristics of the threat model and of the environment being pro-tected. Many of the incidents which have been publicly disclosed in the last years have in fact underlying important facts.

The complexity of the environments is often very difficult to handle. For instance, a nuclear plant in Georgia was shutdown for 48 hours as a consequence of a software update installed on a workstation operating in its business network. Nobody was aware of the connection between the workstation in the business netowkr and the control system on the SCADA network, and of the effects caused by this connection.

Additionally, incidents witnessed in the recent years have underlined an unprecedented level of sophistication in the threats targeting ICS environments. A prime example of this sophistication is the Stuxnet infection, which used *four* distinct and previously unknown zero-day exploits, and leveraged *multiple* stolen certificates for the injection of its rootkit: when the certificate used for the installation of the rootkit was reported stolen, and consequently revoked by the Symantec (Verisign) Certification Authority, the malware was immediately patched remotely to utilize a second stolen certificate. Stuxnet was not an isolated incident: in 2011, a threat sharing similar characteristics to Stuxnet was discovered and was shown to have been generated by the same authors, or those who have access to the Stuxnet source code [7].

These examples highlight specific challenges of protecting critical infrastructure environments, challenges which can be traced back to how critical infrastructure environments differ from typical enterprise environments.

Critical infrastructure environments are very heterogeneous. They include a mix of traditional desktop computers, large mainframes, and field devices. These devices are profoundly different in terms of computational power, communication protocols and even in their ability to be managed and provisioned (i.e. install new software or upgrades). The manner in which these devices are interconnected can vary significantly from company to company (even in the same business branch, such as energy) and automated management controls are frequently non-existent. Because of this heterogeneity in hardware, software and network topology, the security assessment of these environments is particularly challenging. Preliminary studies performed on some of these devices have shown that the security of those systems has been neglected and that a motivated attacker could easily penetrate those systems.

Many communication protocols are vendor-specific. While standards exist for many communications protocols [13], vendors have added specific extensions to provide additional functionalities. The lack of publicly available information on these extensions and their interactions negatively impacts standard security mechanisms, including most Intrusion Detection Systems, which generally rely on signatures for the detection of threats, as well as standard vulnerability discovery tools which require knowledge of the protocol specifications in order to properly assess the robustness of protocol implementations.

Critical infrastructure environments are very valuable targets. Because of their strategic importance, critical infrastructure environments are likely to be targeted by highly motivated and resourceful attackers. The motivation and resources available to individuals interested in compromising these systems can be considerably greater than those attacking more typical IT environments. Many security practices that aim at preventing intrusion by raising their cost (e.g. requiring valid signatures to load kernel drivers) may be ineffective when dealing with these highly resourceful attackers.

2. The CRISALIS project

The CRISALIS project (**C**ritical **I**nfrastructure **S**ecurity **A**nalysis), is a Research Project funded by the European Commission in the context of the FP7 research framework that aims at revisiting the convergence between standard IT systems and industrial control systems typically used in the context of the power grid from a security standpoint. The project will involve in a three-year effort a set of key actors in European academic research (EURECOM, Chalmers University, University of Twente), in the manufacture of devices (Siemens), in the development of security solutions (Symantec) as well as in the deployment and maintenance of national infrastructures (ENEL, the Italian energy provider, and Alliander, key actor in the deployment of Smart Meters in the Netherlands).

The project is articulated over a set of coherent and pragmatic guidelines:

- Focus the attention on real-world, targeted attacks carried out by resourceful attackers against critical infrastructures.
- Address two specific, yet interlinked, environ¬ments, namely the SCADA systems employed in power generation and distribution as well as the AMI infrastructure employed in in the distribution of electricity to consumers.
- Develop practical solutions and tools, and test these tools on real systems.

The project research effort spans over three main themes: **(i) securing the systems,** by means of novel automated analysis of CI environments and discovery of new threat vectors; **(ii) detecting the intrusions,** by developing new technologies aiming at coping with the heterogeneity of protocols, interactions and devices typical of these systems; **(iii) analyzing successful intrusions,** by devising techniques to facilitate the "post-mortem" analysis of the environments and of specific devices. These three research themes are meant to complement each other in order to address their respective limitations. Not all the security problems can be easily discovered and fixed, therefore the need to improve our capability of detecting anomalous interactions in these environments. Similarly, not all the intrusions can be detected with certainty, therefore the need to develop tools to inspect specific devices and detect tampering. Key to the uniqueness of the project is a specific focus on two main challenges that characterize these environments:

System complexity. An important transversal research theme consists in the development of tools and techniques for the automated discovery of the structure and the interactions in these environments. This will include protocol learning techniques to address the proliferation of vendor-specific protocol dialects in these networks, as well as new device fingerprinting approaches able to deal with the diversity in the involved devices.

Validation. In order to guarantee the practicality and the general applicability of the developed tools and techniques, the CRISALIS project will setup a set of real-world environments for their validation and evaluation thanks to the involvement of critical infrastructure maintainers such as ENEL and Alliander.

The project, set to start on the 1st of March 2012, will ultimately contribute to addressing the current imbalance between challenges and opportunities associated to the convergence of IT technologies with Industrial Control Systems. If accepted, the presentation will present more in depth the research content, the planned methodology and early results of the project.

References

[1] J. Weiss, "Control systems cyber security—the current status of cyber security of critical infrastructures." Testimony before the Committee on Commerce, Science, and Transportation U.S. Senate, March 2009.

[2] H. Luiijf, "SCADA good practices for the dutch drinking water sector," TNO DV 2008 C096, March 2008.

[3] Symantec, "2010 critical infrastructure protection study," October 2010. [Online]. Available: http://bit.ly/bka8UF

[4] ——, "The Stuxnet worm." [Online]. Available: http://go.symantec.com/stuxnet

[5] ——, "Stuxnet: a breakthrough." [Online]. Available: http://www.symantec.com/connect/blogs/ stuxnet-breakthrough

[6] ——, "W32.Stuxnet variants." [Online]. Available: http://www.symantec.com/connect/blogs/ w32stuxnet-variants

[7] ——, "W32.Duqu, the precursor to the next Stuxnet." [Online]. Available: http://go.symantec.com/duqu

[8] J. Weiss, "Control Systems Cyber Security -The Current Status of Cyber Security of Critical Infrastructures," Testimony before the Committee on Commerce, Science, and Transportation U.S. Senate, March 2009.

[9] F. Cohen, "The smarter grid," in *Proceedings of IEEE Symposium on Security and Privacy*, vol. 8, 2010, pp. 60–63.

[10] The FORWARD FP7 project, "Forward whitebook," January 2010. [Online]. Available: http://www.ict-forward.eu/media/publications/ forward-whitebook.pdf

[11] M. Davis, "Recoverable advanced metering infrastructure," Blackhat USA, 2009.

[12] R. Anderson and S. Fuloria, "Who controls the off switch?" in *IEEE International Conference on Smart Grid Communications*, October 2010.

[13] "Opc foundation." [Online]. Available: http://www. opcfoundation.org/

Challenges in Critical Infrastructure Security

Corrado Leita

Symantec Research Labs

1

Symantec Research Labs

- CARD (Collaborative Advanced Research Department) group
  - Sophia Antipolis, FR
  - Culver City, CA
  - Herndon, VA
- Relevant recent work:
  - **SGNET**: distributed honeypot deployment for the study of code injection attacks based on ScriptGen
  - **HARMUR:** dataset providing a historical perspective on client-side threats
  - **TRIAGE:** multi-criteria decision analysis for the study of security datasets (Olivier Thonnard)
  - **WINE:** Worldwide Intelligence Network Environment (http://www.symantec.com/WINE)

✓ Symantec. 2

Convergence between IT and OT technologies

- Interconnection of standard computer systems with industrial control systems

- An **opportunity**?
  - Lower costs and increased system efficiency
  - Opportunity to leverage standard IT techniques (intrusion detection, file scanning, standard hardening techniques, ...)
  - Opportunity to enable OT suppliers to manage and support OT devices at scale

- A **threat**?
  - Enable attacks and incidents that are typical of standard IT environments
  - Enable attacks on critical infrastructures and environments such as energy, gas, medical
  - Privacy violations from data being more widely available

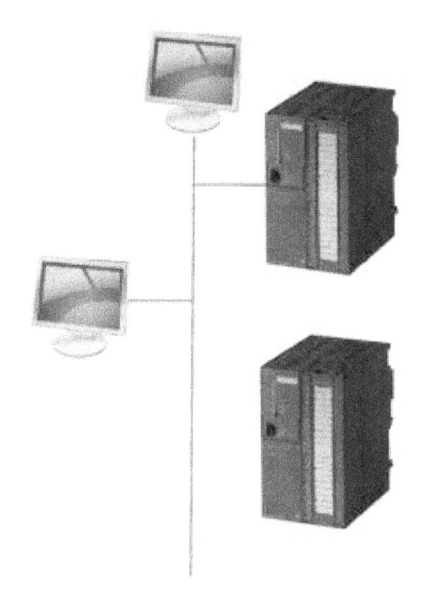

Cybersecurity in CPS Workshop - CRISALIS FP7 project

 Symantec. 3

What are the challenges in the protection of ICS environments?

Cybersecurity in CPS Workshop - CRISALIS FP7 project

 Symantec. 4

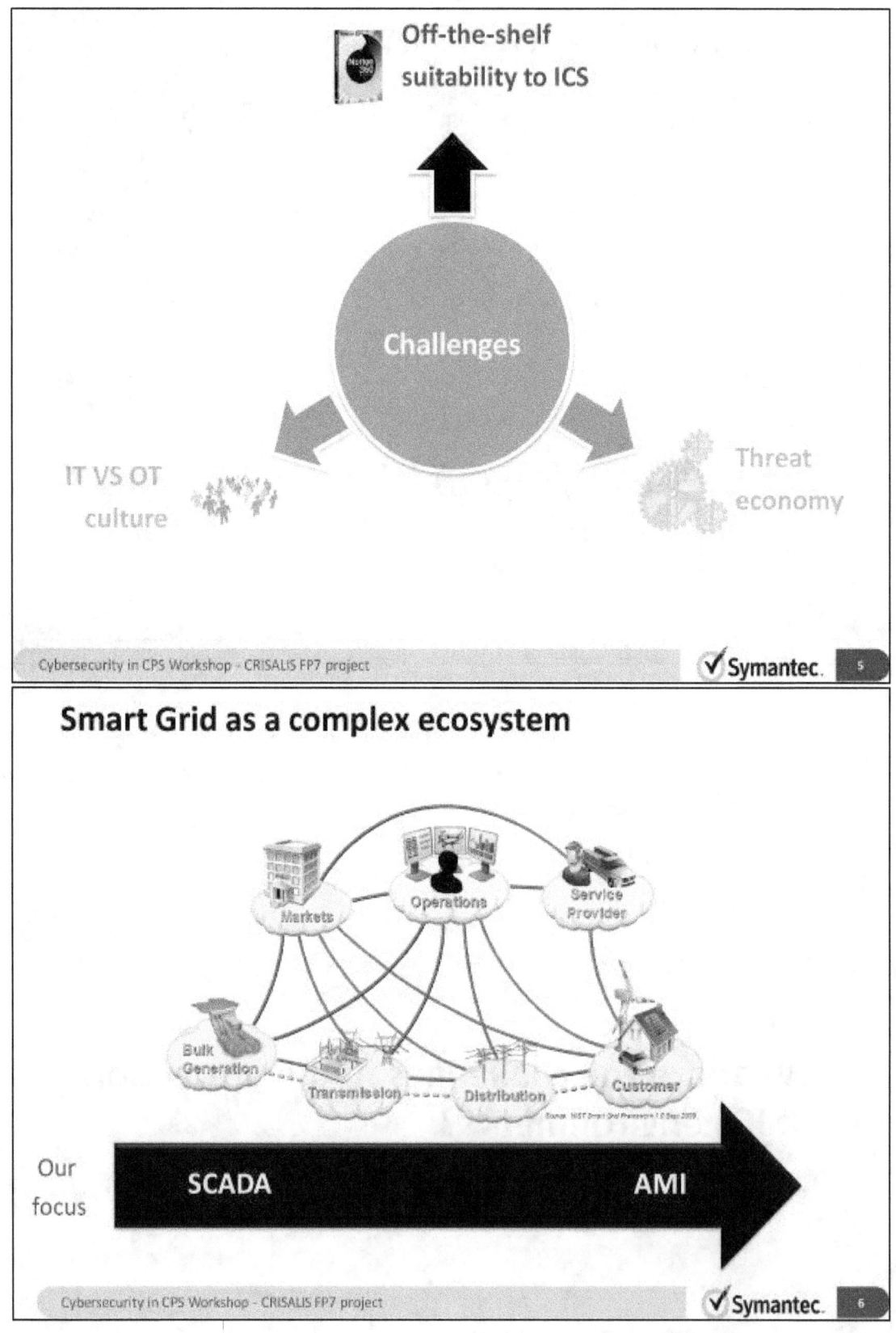

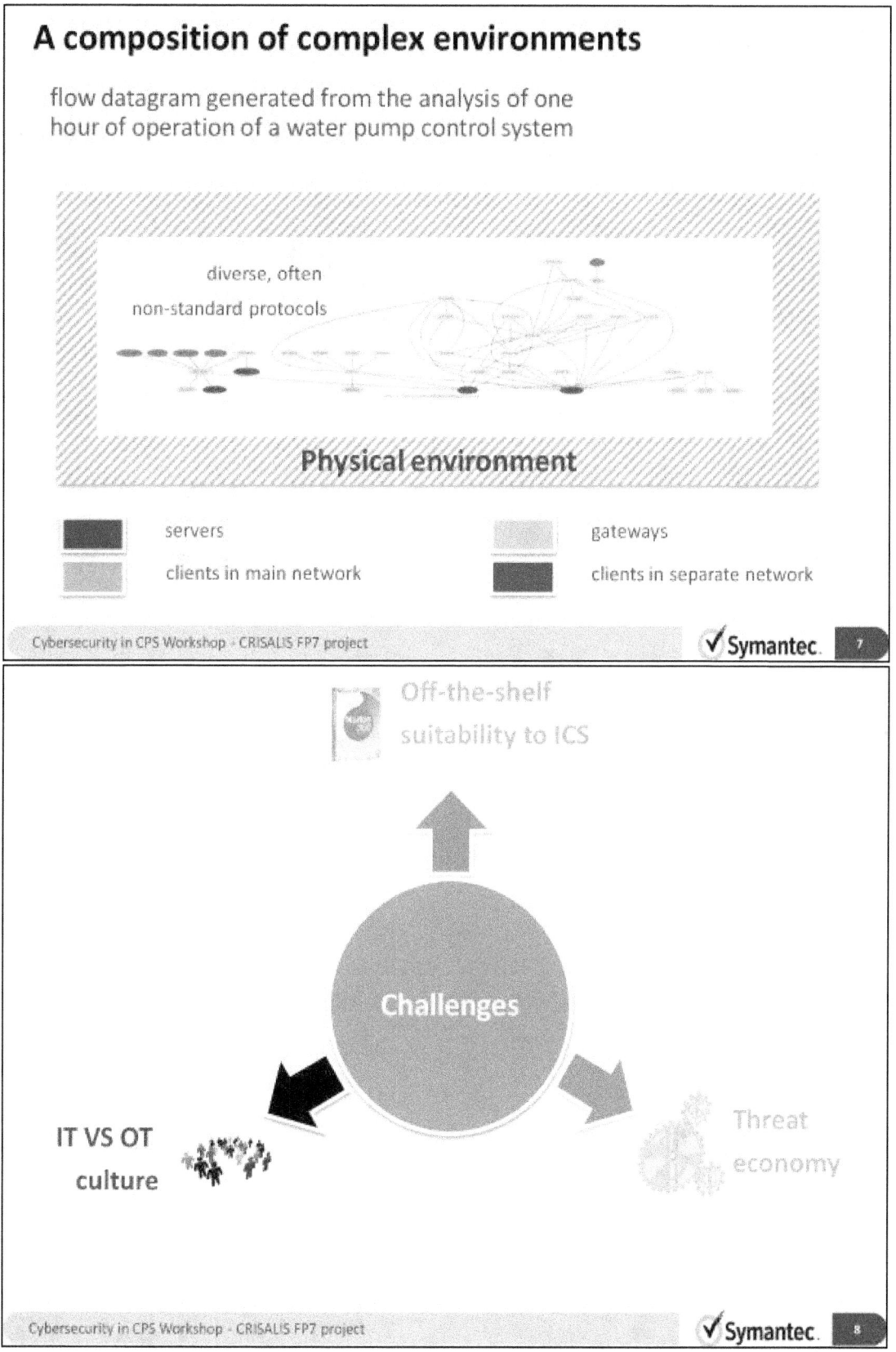

The interesting lesson

Is it possible to burn-out a water pump by solely interfacing with the SCADA layer? Fail-safe mechanisms exist to prevent physical damage!

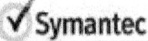

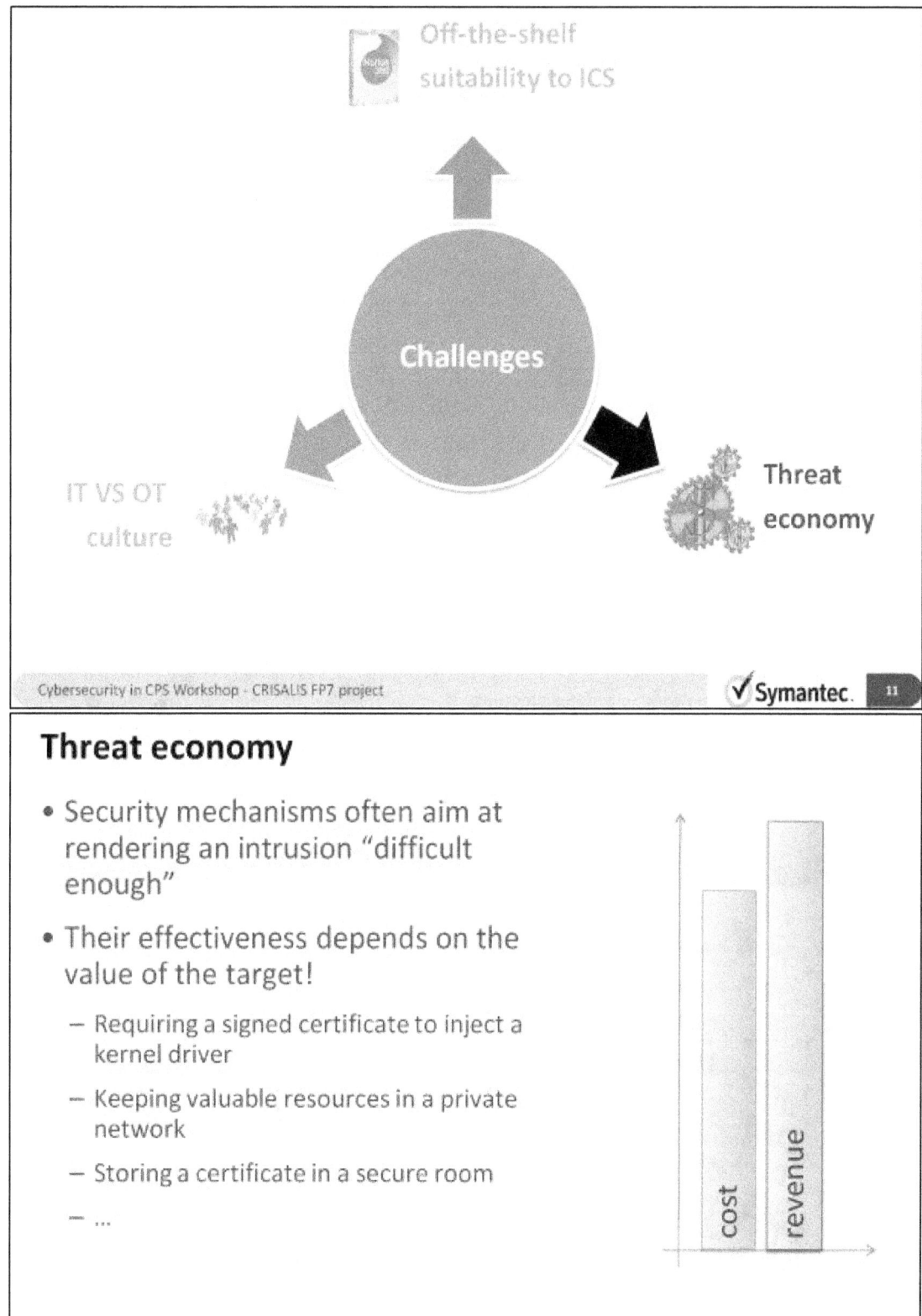

The threats are real

Symantec 13

What is your experience with each of this type of attacks? (1580 industries contacted, 2010)

Symantec 2010 Critical Infrastructure Protection Study - **http://bit.ly/bka8UF**

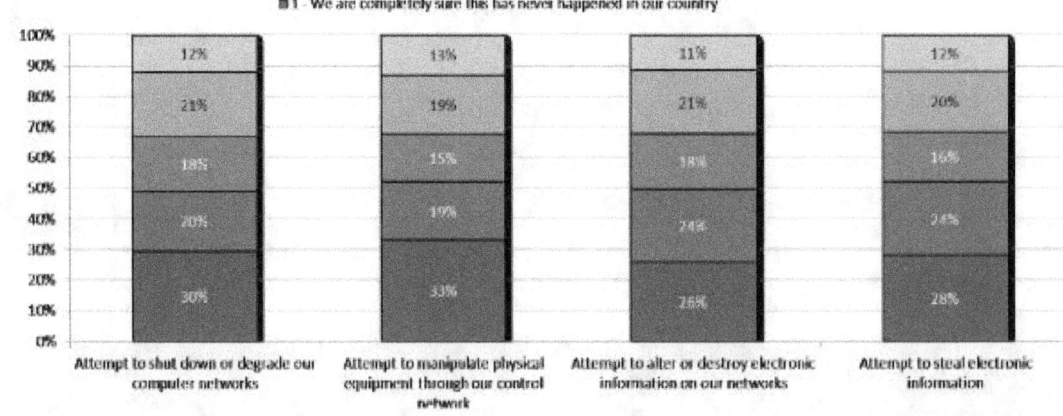

Symantec 14

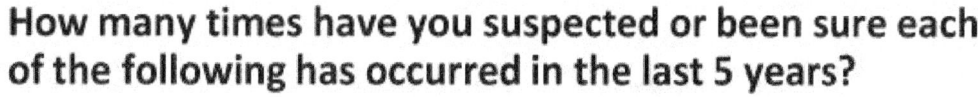

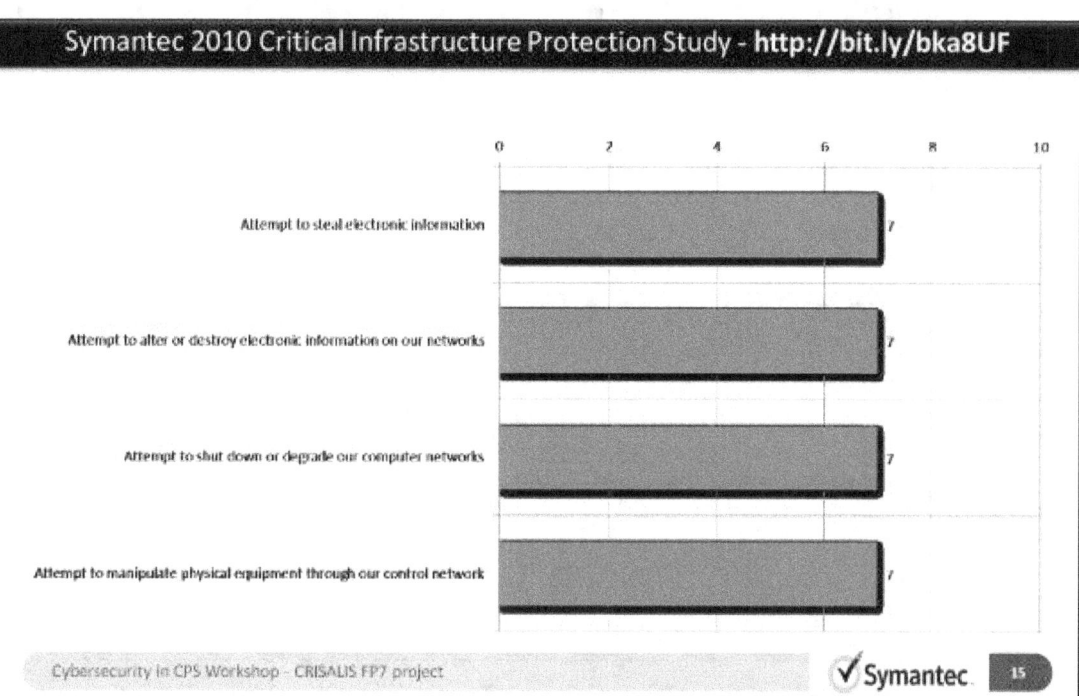

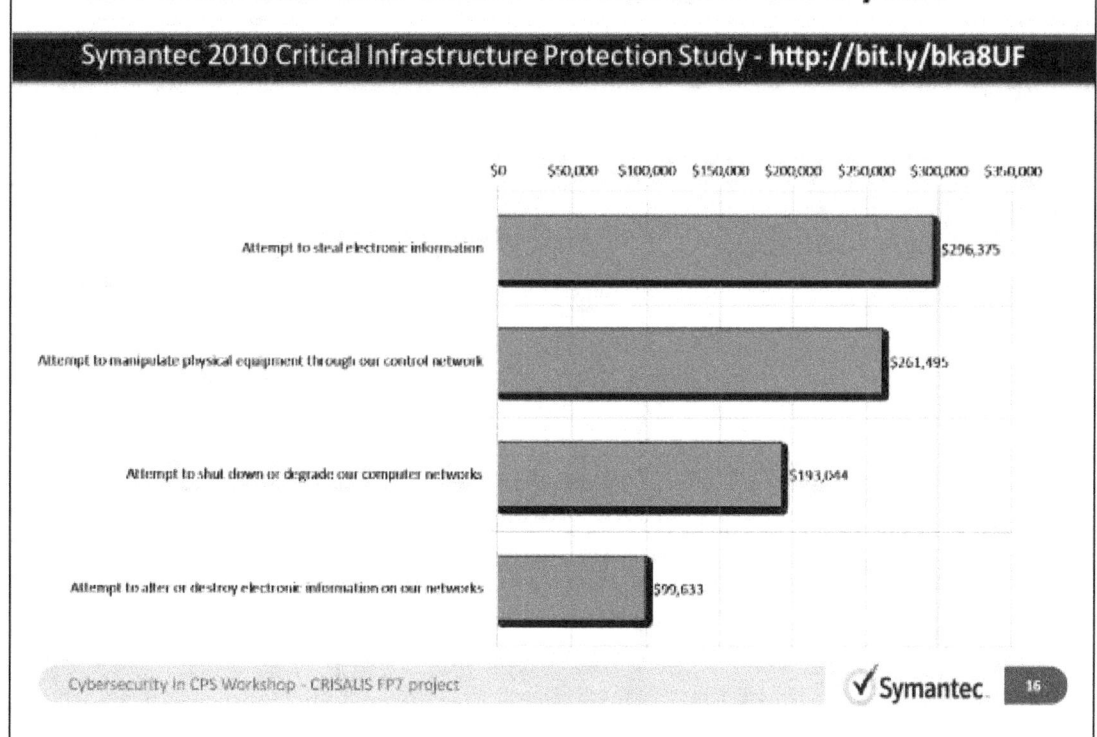

Stuxnet

- Windows worm discovered in **July 2010**

- Uses **7** different self-propagation methods

- Uses **4** Microsoft 0-day exploits + **1** known vulnerability

- Leverages 2 Siemens security issues

- Contains a Windows rootkit

- Used **2 stolen digital certificates** (second one introduced when first one was revoked)

- Modified code on Programmable Logic Controllers (PLCs)

- First known PLC rootkit

Cybersecurity in CPS Workshop - CRISALIS FP7 project

✓ Symantec.   17

Stuxnet and the myth of the private network

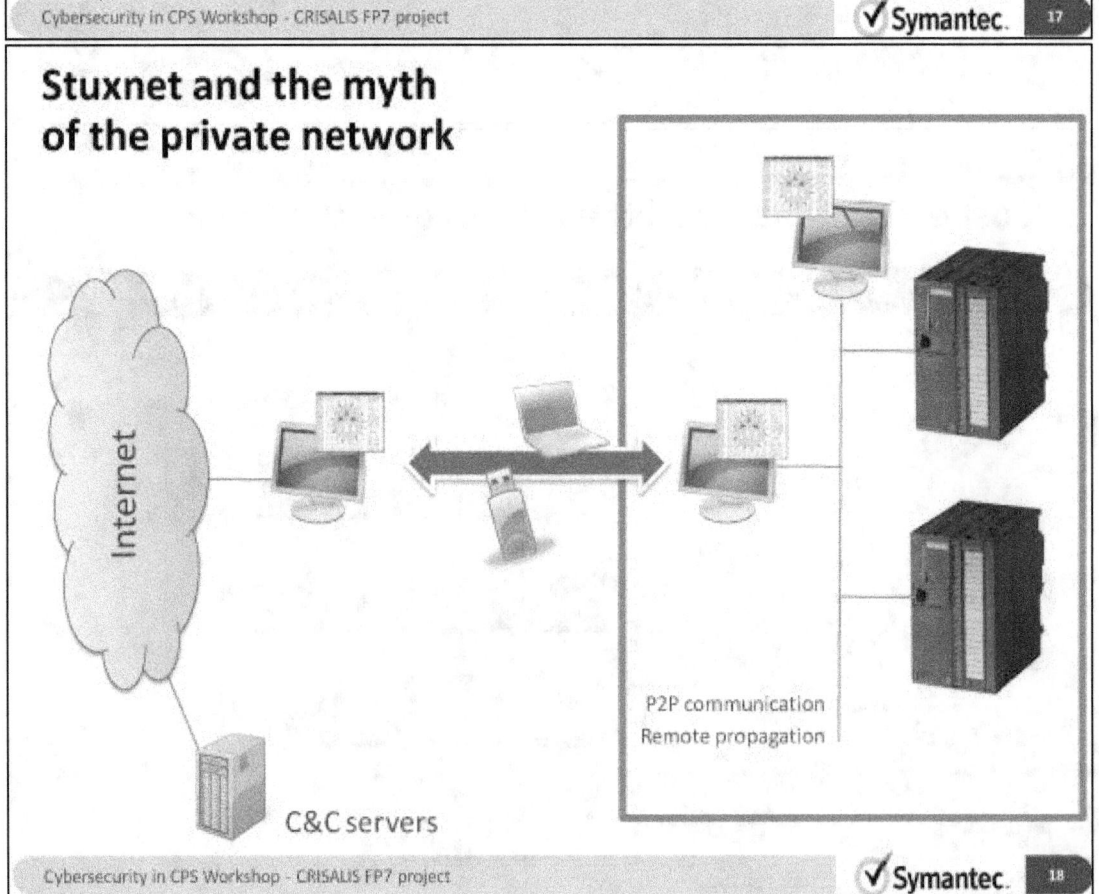

Internet

C&C servers

P2P communication
Remote propagation

Cybersecurity in CPS Workshop - CRISALIS FP7 project

✓ Symantec.   18

Stuxnet: an isolated incident?

- **September 2011:** a European company seeks help to investigate a security incident that happened in their IT system, and contacts CrySyS labs (Budapest University of Technology and Economics)

- **October 2011:** CrySyS labs identifies the infection and shares information with major security companies
  - Duqu: named after the filenames created by the infection, starting with the string "~DQ"
  - A few days later, Symantec releases the first report on Duqu malware sample with the help of the outcomes of the original CrySyS investigators

✓Symantec. 19

Extremely stealthy and targeted infection

Infection leaves almost no trace on hard drive: only the driver file is stored in stable storage!

- 0-day vulnerability in TTF font parser

- Shellcode ensures infection only in an 8 days window in August

- No self-propagation, but spreading can be directed to other computers through C&C
  - Secondary target do not communicate with C&C, communicate instead through P2P

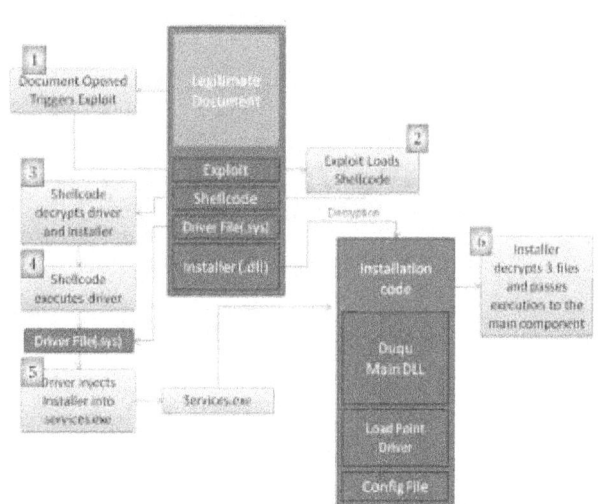

✓Symantec. 20

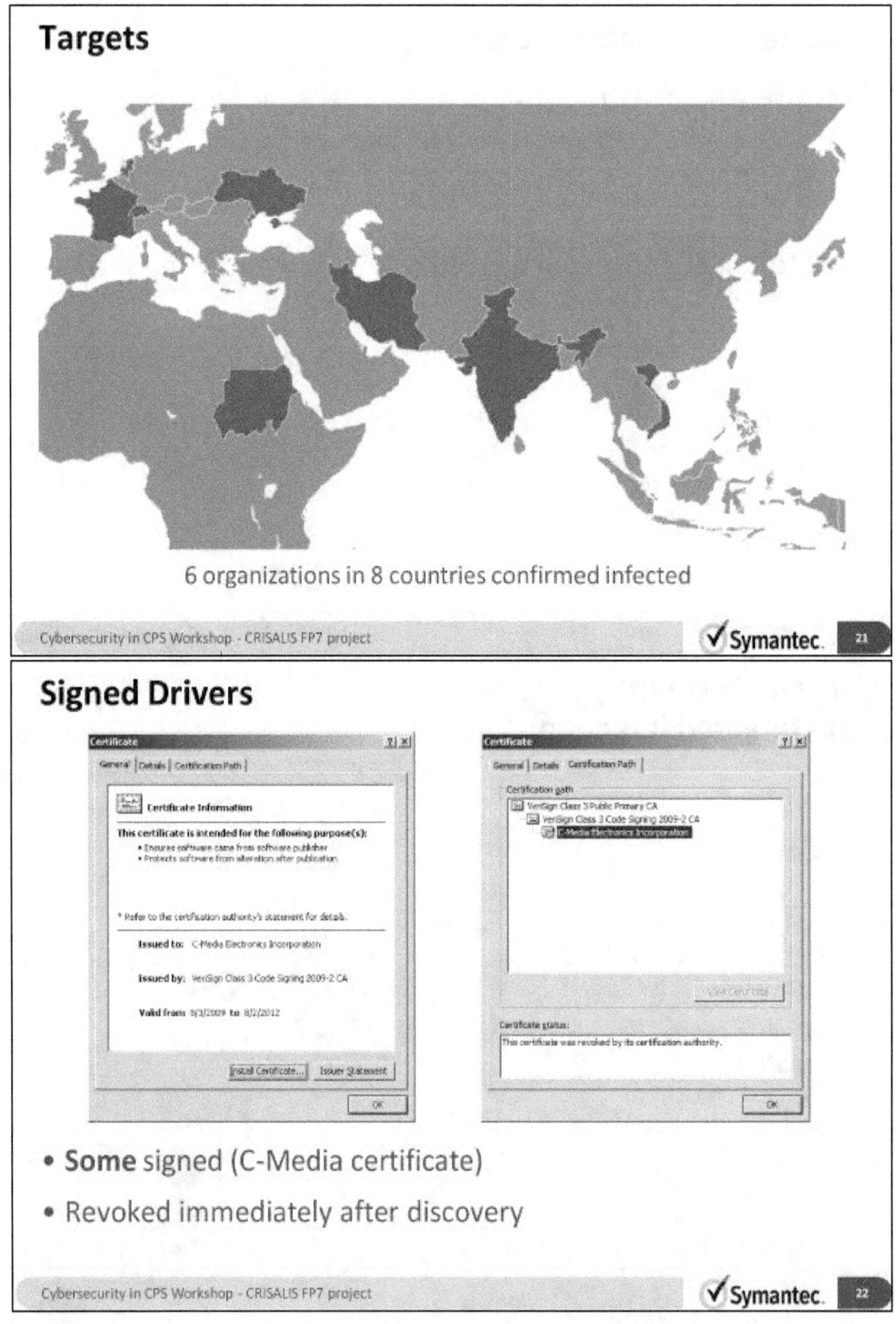

Command & Control Complexity

- Communication over TCP/80 and TCP/443
  - Embeds protocol under HTTP, but not HTTPS
  - Includes small blank JPEG in all communications
  - Basic proxy support
- Complex protocol
  - TCP-like with fragments, sequence and ack. numbers, etc.
  - Encryption AES-CBC with fixed Key
  - Compression LZO
  - Extra custom compression layer
- CnC server hidden behind a long sequence of proxies

✓Symantec. 23

Duqu "strange clues"

- TTF Exploit
  - Font name "Dexter Regular" from "Showtime Inc."
  - Only two characters defined:

    :)

- Inside the keylogger component is a partial image
  - "interacting Galaxy System NGC 6745"

✓Symantec. 24

Stuxnet and Duqu

- Stuxnet: first publicly known malware to cause public damage

- Duqu: shares many similarities, used for cyber espionage (a new Stuxnet?)

- High complexity
  - Require resources at the level of a nation-state
  - The attackers are not gone: new binary found compiled in February 2012

- **Cyber warfare is not a myth**

Symantec 25

CRISALIS

Symantec 26

What have we learned so far?

1. **Attacker motivation:** no security practice is likely to make the intrusion **difficult enough**. New motivations for attackers (crime, cyber warfare) mean more resources and incentives to conduct attacks.

2. **Myth of the private network:** also because of 1. , relying on network isolation from the Internet as main security protection is ineffective. Physical security cannot be enforced in practice, and network isolation renders cloud-based security technologies impossible to apply (e.g. reputation, data analysis, signatures, …).

3. **From Intrusion Prevention to Intrusion Tolerance:** a layered approach is required with several safety nets and managerial procedures to handle fallback modes.

Cybersecurity in CPS Workshop - CRISALIS FP7 project  27

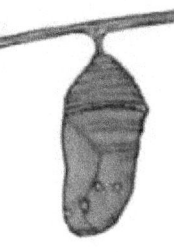

The CRISALIS project

- 3-year collaborative project (funded by FP7-SEC)
- Participants:

– Symantec (Ireland)
– Siemens (Germany)
– Security Matters (Netherlands)
– EURECOM (France)
– Chalmers (Sweden)
– University of Twente (Netherlands)
– ENEL (Italy)
– Alliander (Netherlands)

Industry
Academia
End users

Cybersecurity in CPS Workshop - CRISALIS FP7 project  28

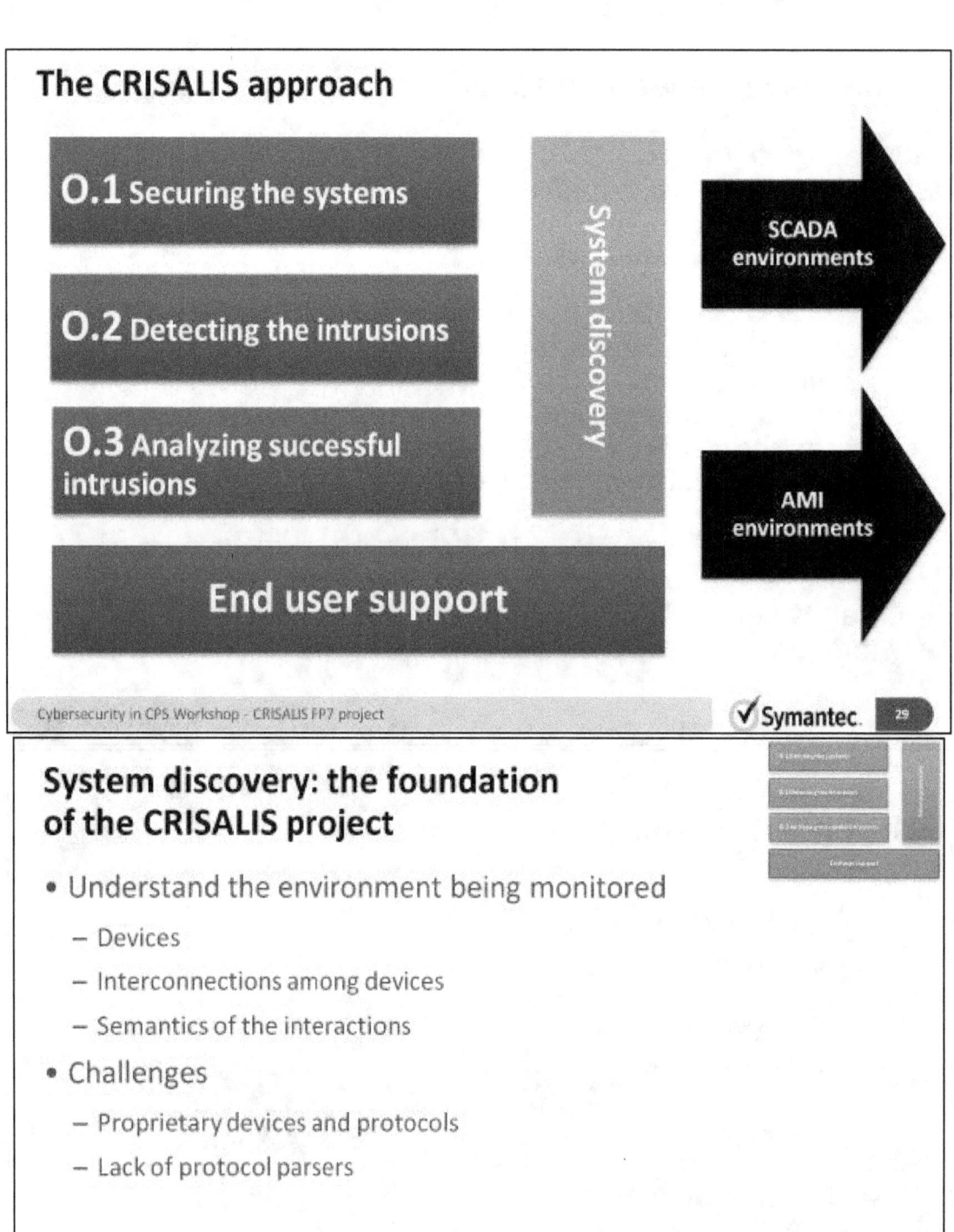

The CRISALIS approach

O.1 Securing the systems

O.2 Detecting the intrusions

O.3 Analyzing successful intrusions

System discovery

End user support

SCADA environments

AMI environments

Cybersecurity in CPS Workshop - CRISALIS FP7 project — Symantec — 29

System discovery: the foundation of the CRISALIS project

- Understand the environment being monitored
  - Devices
  - Interconnections among devices
  - Semantics of the interactions
- Challenges
  - Proprietary devices and protocols
  - Lack of protocol parsers

Cybersecurity in CPS Workshop - CRISALIS FP7 project — Symantec — 30

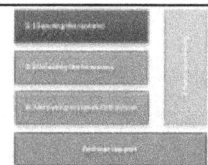

O.1 Securing the systems

- Penetration testing
  - Globally accepted methodologies in ICT infrastructures
  - Methodology needs to be carefully revisited to be applicable to ICS (dangerous!)
- Vulnerability discovery
  - Attention to the **automated** discovery of vulnerabilities in ICS devices
    - Static analysis of the binary code
    - Dynamic analysis
  - Drive the vulnerability discovery process through information on the protocol specification

 Symantec 31

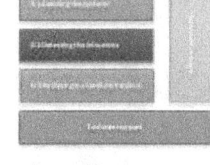

O.2 Detecting the intrusions

- Vulnerability discovery is unlikely to exhaustively identify all the possible threat vectors. How to identify and block a successful intrusions?
- Targeted attacks: we need to avoid a-priori assumptions on the threat vector
  - Traditional assumptions on the threat model are likely to not hold
  - Signature-based technologies are not appropriate
  - Revisit behavior-based detection in ICS environments
  - Revisit host-based monitoring techniques

O.3 Analyzing successful intrusions

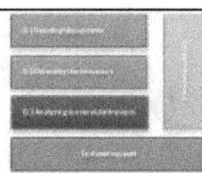

- Be ready to fail: provide instruments to detect suspicious modifications to the devices and analyze their effects
  - Forensic analysis of industrial devices: how can we understand if a PLC device has been compromised? How can we understand the impact of the modifications?
- Challenges
  - Perceived absence of real threats by the industry
  - Deployment of proprietary components and protocols
  - Lack of persistent storage capabilities

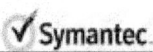

Cybersecurity in CPS Workshop - CRISALIS FP7 project — Symantec 33

Validation environment

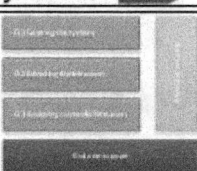

- How can we validate the soundness of the obtained results? What is the performance of an intrusion detection methodology in real world environments?
- Validation environments:
  - ENEL Security Lab (Livorno, Italy): replica of a real-world SCADA system used in power generation
  - Alliander Testing deployment (Netherlands): testing AMI deployment

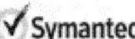

Cybersecurity in CPS Workshop - CRISALIS FP7 project — Symantec 34

An example: CRISALIS and protocol learning

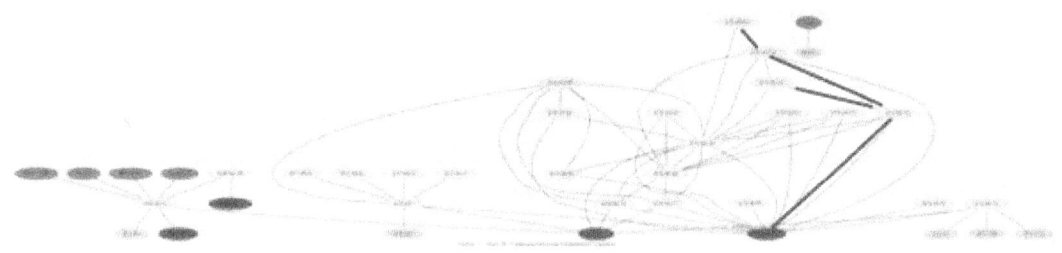

- Can we try to attach semantics to the different edges with no a-priori knowledge on the protocol structure?

- Can we infer causality...?

✓Symantec 35

ScriptGen

- Protocol-agnostic algorithm
- Observe conversation samples between a client and a real server
- Infer semantics using bioinformatics algorithms
- Proved good results in handling deterministic exploit scripts

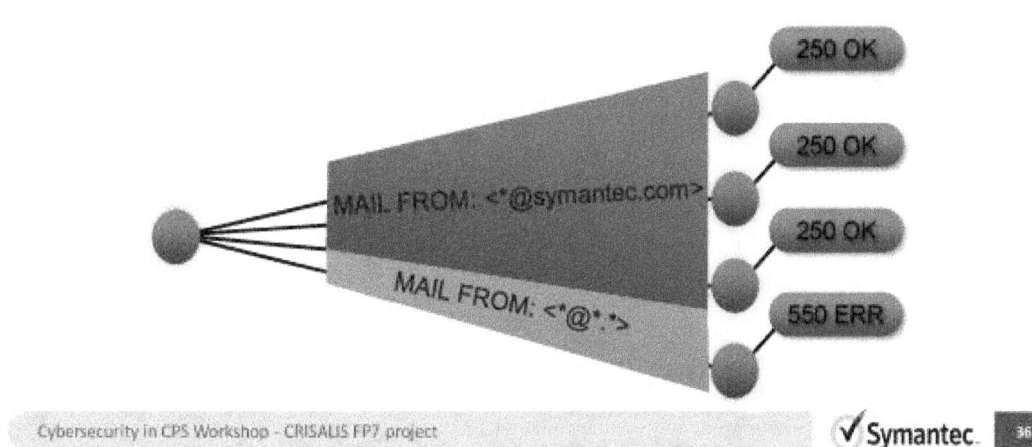

✓Symantec 36

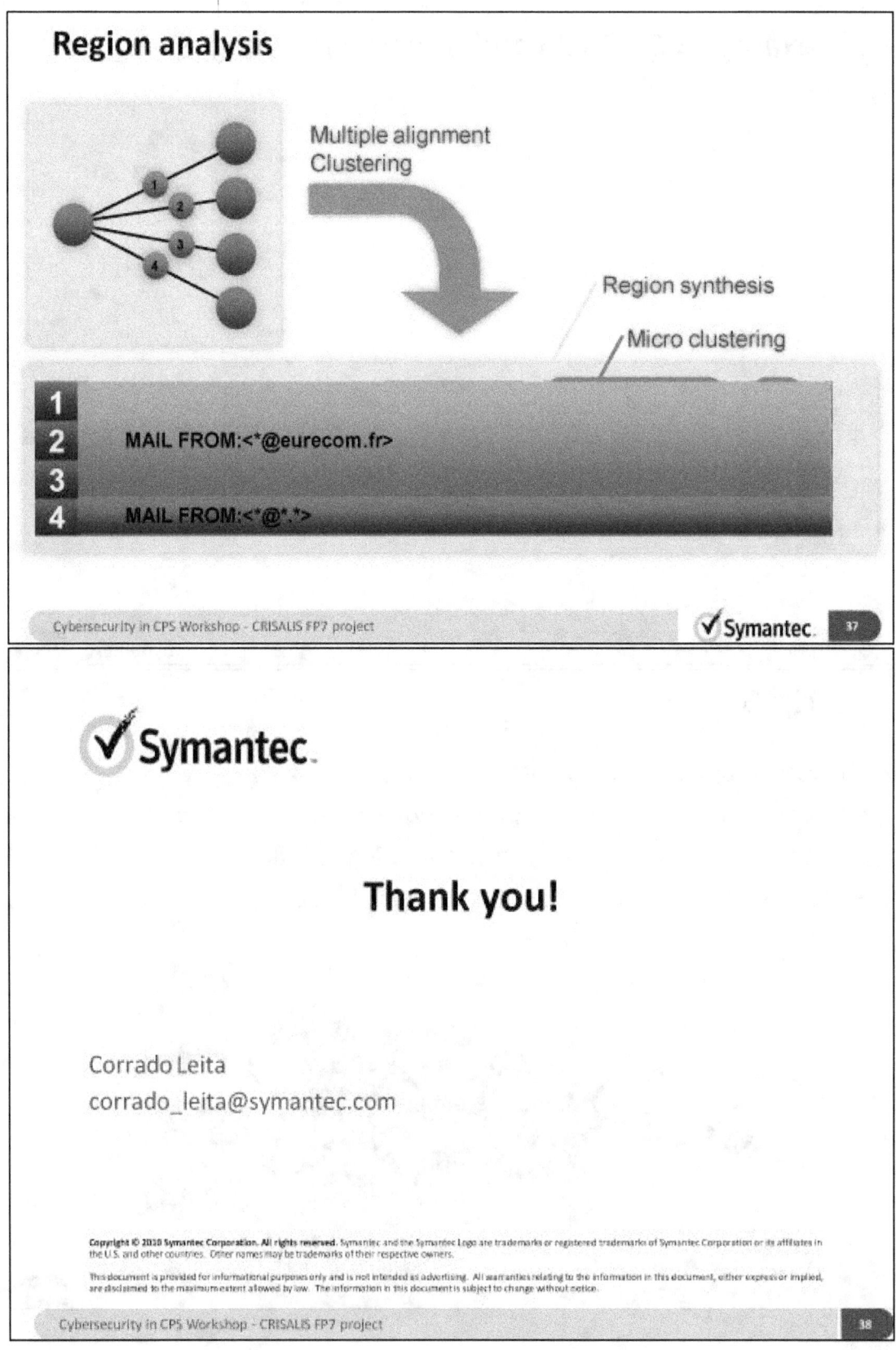

11. Understanding the Role of Automated Response Actions to Improve AMI Resiliency

Ahmed Fawaz, Robin Berthier, William H. Sanders
Information Trust Institute and Department of Electrical
& Computer Engineering
University of Illinois at Urbana-Champaign
Urbana, IL, USA
{afawaz2, rgb, whs}@illinois.edu

Partha Pal
BBN Technologies
Boston, MA, USA
ppal@bbn.com

Abstract—The smart grid promises better services and higher reliability but is exposed to new security threats. In particular, deployment of advanced metering infrastructure (AMI) will vastly increase the attack surface because of the smart meters installed in customer homes. Managing the security of AMI cannot be done manually because of its size and complexity. Thus, we propose a three-step plan to bring automated responses to AMI. Considering the challenges of automated responses, we will develop a taxonomy of response actions in AMI. Then, we will model the response actions in terms of their impact and cost for the different actors in the system: customers, administrators, and attackers. Finally, we will discuss implementation and evaluation requirements for a practical automated response engine for AMI.

Keywords: AMI, CPS, Response action, Cyber security

The adoption and deployment of the smart grid promise customers faster and more reliable service. The smart grid enables those improvements through its capabilities for remote control, instant detection of blackouts, and accurate state estimation of the power grid using phasor measurement units (PMU). Additionally, the smart grid accommodates more customer services, such as real-time pricing, and includes provisions for future electrical vehicles. A core component of the smart grid is the advanced metering infrastructure (AMI). An AMI is the communication solution for smart meters that transmit real-time meter readings to the administrative network and receive commands to control service remotely. An AMI enables fine-grained detection of blackouts and will thus enable faster customer service.

A typical AMI will allow remote control of every smart meter, including the ability to turn service off, and in some scenarios will allow utility companies to control specific appliances in individual homes as part of environmental programs that offer reduced prices at certain hours of the day. Moreover, utility companies will no longer need to have human meter readers drive around and obtain monthly readings, because readings will be sent to the utility company frequently from the meters through the AMI network. Finally, the introduction of smart appliances that can communicate with smart meters to get realtime pricing information means that owners will be able to control those appliances remotely via the Internet.

AMI presents more security threats than regular cyber-physical systems (CPS) do, as its architecture and services allow for a larger attack surface. The attack surface includes 1)

the corporate network, 2) the wireless mesh network, 3) the home area network, and 4) meters that are within the reach of customers. Possible threats can be classified according to attack scale, ranging from relatively small-scale targeting of specific houses (in order to turn off service or specific appliances, such as alarm systems) or stealing of energy (through alteration of meter readings or duplication of meters), up to large organized crimes that target large geographical regions. Moreover, attacks could target the control commands sent by a utility to its AMI. Additional security issues also rise from the use of the wireless technology for smart meter communication. Additional attacks will be facilitated by the wireless mesh network that will be used to connect meters; such networks are prone to single points of failure, availability problems, jamming, eavesdropping, man-in-the-middle attacks, and wormhole and black hole attacks [1,3,5].

Compared to traditional IT systems, AMI has stringent requirements in terms of quality of service and security guarantees. Those requirements include:

1. Availability: Utility companies should be able to get the latest meter readings and send out control commands within specific time constraints. Moreover, customers expect the latest pricing to be available.

2. Resilience: AMI provides a critical service to customers. It must be able to work under extreme conditions and provide the core service of measuring energy consumption even under attack.

3. Fast recovery: In the event of an attack, a compromise, equipment faults, or even blackouts, an AMI should allow fast recovery and restoration of service.

4. Size: In the future, a typical AMI could be larger than any conventional CPS ever built , with millions of nodes in cities; this massive size imposes scalability issues for traditional security solutions.

5. Privacy: There are also privacy concerns specific to AMI, since the readings and commands sent between the meter and the utility company reveal private information about customers.

Important efforts (by researchers and by organizations such as NERC and NIST) have been made to promote security solutions for AMI networks, such as VPNs, encryption [4], and remote attestation [8]. Such efforts are important, but cannot completely secure systems, mainly because vulnerabilities can always be found in the implementations of protocols and applications, or in the human operators who can provide access to restricted resources unintentionally. Moreover, since meters are left without real physical protection, tampering with devices may leak secret keys stored in internal memory and thus cause security breaches in the network. Consequently, traditional attack prevention solutions have to be supplemented with detection and mitigation approaches. Our present work focuses on studying the possibility of a framework that can automatically respond to cyber intrusions given the requirements of an AMI.

The importance of intrusion detection for AMI is still critical, and several approaches have been proposed [2,11]. However, intrusion detection is prone to inaccuracies, and monitoring such a large number of nodes will rapidly lead to an unmanageable volume of alerts and demands for decisions. The combination of potentially weak detection capabilities and stringent CPS requirements means that to offer strong resiliency against cyber-attacks, security solutions have to be proactive. For example, the uncertain identification of a suspicious behavior has to trigger the automated deployment of additional monitoring capabilities to translate inaccurate reports into actionable information. A variety of automated response solutions have been studied over the past decade [13], but none have been tailored for the specific requirements of complex cyber-physical systems such as AMI. Moreover, the practicality of existing solutions is limited, and for multiple reasons, the industry has been reluctant to implement sophisticated automated response actions. First, implemented actions are often all-or-nothing, meaning that they lack the flexibility to adapt to various situations and can lead to dramatic consequences in the case of false positives. Second, there is a poor understanding of the impact of response actions in large and complex CPS. Third, that lack of understanding can result in vulnerabilities in the response action itself, which could enable attackers to game the system and cause automation to do more harm than good.

We gained a better understanding of the limitations of current automated response solutions by reviewing related work from the perspective of practicality for the specific requirements of AMI. As a result, we plan to present the following approach to bringing efficient and secure automated response to AMI.

The first step, which is in progress, involves development of a taxonomy of response actions that suits AMI requirements, such as always preserving the mission of delivering energy and accurately measuring consumption. The taxonomy will allow us to construct a set of possible response actions by emphasizing the concept of flexibility. Flexible actions can be tuned to meet a wide variety of requirements and situations. This will then guide the development of a practical case study to design flexible actions for an AMI. The taxonomy has two high-level categories: 1) learning actions, and 2) modifying actions. Learning actions are either passive or active and are designed to gather additional information about security incidents. Learning actions include enabling of additional IDS sensors with a higher granularity, logging of traffic, or active sending of probe packets to locate compromised nodes. Modifying actions work to respond to and recover from an attack. Modifying actions have two subcategories: recovery actions and limiting actions. Limiting actions reduce privileges of a given entity, thus reducing its ability to propagate an attack. Limiting actions include addition of firewall rules to block a meter's traffic, changes to access privileges to certain resources within a meter, and changes to routes within the mesh network to avoid a compromised meter. Recovery actions will work to stop attacks and return to a previous working state in the system; such responses include application of update patches, flashing of a clean OS version, and even sending of field technicians to change a meter.

The second step after building the taxonomy will be to model the response actions' impact and cost. The first task in modeling response actions will be to identify the different actors in our system. Usually, security researchers consider the main actors to be the attacker and the administrator. However, we propose to include customers as well, since they can also be affected by the attacker's actions and the administrator's reactions. An action's impact can be described as good or malicious, where good actions are those that benefit legitimate entities (administrators and customers) and negatively impact illegitimate entities (i.e., attackers) by making it harder for them to achieve their malicious goals. In order to quantify the impact of an action, it is necessary to define the cost of the action. Several researchers have proposed methods to compute the cost of actions [6,7,10]; some use the difference in the value of a security metric based on dependency graphs between the system states before and after an action was done. Others decompose the cost based on the number of unavailable resources, impact on the system, and operation cost. Most research uses a weight matrix for the different confidentiality, integrity, and availability (CIA) metrics to describe the importance of each security property. Most previous work does not look into practical ways to compute the cost of an action to the customer, or consider the time needed to recover as part of the cost. Moreover, the use of a static matrix to specify the importance of each security property is highly subjective and does not provide a method to compute those values based on the policies of the corporation, or even provide a sense of how to tweak the values to change the reactions of the system. Additionally, cost assessment in the context of a CPS requires a detailed understanding of the interfaces between the cyber and the physical mechanisms. Because of those limitations, we will propose a cost computation method that allows us to consider the cost for customers. It will also allow for flexible cost for actions with varying intensity (e.g., rate limiting with a variable threshold rate). Moreover, we will propose clear methods to generate and tweak the weight coefficients needed to compute the cost of an action, as well as include the impact on physical systems in the calculation.

The next step in this project will be to explore solutions for automatic selection of response actions at runtime during an attack. We plan to study the game-theoretic response and recovery engine (RRE) proposed by Zonouz et al. [12]. RRE models the system as a Stackelberg game [9] between the attacker and the administrator. RRE uses an attack-response tree to represent the possible attacker moves and tags each move with a set of possible responses. Upon an attacker's move, RRE then computes an optimal strategy for the current security state of the system that maximizes the benefit for the administrator while reducing the benefit for the attacker. Several challenges must be addressed before that type of online automated decision-makers become "AMI-ready." First of all, because of the large size of an AMI, the attack-response tree representing the system will get much larger than those of traditional networks, making it difficult for RRE to compute the optimization. Thus, we need an abstraction to reduce the search space of RRE for the AMI. The main idea behind the abstraction is to use the hierarchy within AMI, so we willdivide the attack goals into several interim goals that can be solved independently within a neighborhood. Then, we will form another tree that combines several neighborhoods and decides on high-level actions (e.g., isolating a complete neighborhood). Moreover, RRE does not have provisions for customer costs, and changes are needed to include those costs as part of computing the optimal response strategy.

The final contribution of this project will be to discuss how to evaluate the framework in a realistic environment. We will present a set of experiments that we plan to implement in the TCIPG/Itron testbed. This testbed emulates hundreds of virtualized meters combined with hardware meters, all clustered into several neighborhoods (reflecting a realistic AMI). Each cluster has a collector that sends the readings back to the head end or sends commands from the head end to the meters.

This paper presents a rigorous research plan to study automated response within the unique requirements of an AMI. The proposed solution will help utilities improve on services, operation costs, and reliability. Automated responses in AMI will reduce the maintenance cost for utilities, as it improves the ability to troubleshoot the distribution network by providing situational awareness. Moreover, automated response has the potential to significantly reduce the load on human operators by automatically managing low-level alarms generated by sensors in the network.

REFERENCES

[1] M. Al-Shurman, S. M. Yoo, and S. Park, "Black hole attack in mobile ad hoc networks," in Proceedings of the 42nd Annual Southeast Regional Conference, 2004, pp. 96-97.

[2] R. Berthier, W. H. Sanders, and H. Khurana, "Intrusion detection for advanced metering infrastructures: Requirements and architectural directions," in Proc. First IEEE International Conference on Smart Grid Communications (SmartGridComm), 2010, pp. 350-355.

[3] L. Buttyan and J.-P. Hubaux, Security and Cooperation in Wireless Networks. Cambridge University Press, 2007.

[4] A. Hahn and M. Govindarasu, "Cyber attack exposure evaluation framework for the smart grid," IEEE Transactions on Smart Grid, vol. 2, pp. 835-843, 2011.

[5] Y. C. Hu, A. Perrig, and D. B. Johnson, "Wormhole attacks in wireless networks," IEEE Journal on Selected Areas in Communications, vol. 24, pp. 370-380, 2006.

[6] W. Kanoun, N. Cuppens-Boulahia, F. Cuppens, S. Dubus, and A. Martin, "Success likelihood of ongoing attacks for intrusion detection and response systems," in Proc. International Conference on Computational Science and Engineering (CSE'09), ,2009, pp. 83-91.

[7] N. Kheir, H. Debar, N. Cuppens-Boulahia, F. Cuppens, and J. Viinikka,"Cost evaluation for intrusion response using dependency graphs," in Proc. International Conference on Network and Service Security, (N2S'09), 2009, pp. 1-6.

[8] M. LeMay and C. Gunter, "Cumulative attestation kernels for embedded systems," in Proceedings of the 14th European conference on Research in computer security (ESORICS 2009), pp. 655-670, 2009.

[9] P. Paruchuri, J. P. Pearce, J. Marecki, M. Tambe, F. Ordonez, and S. Kraus, "Playing games for security: An efficient exact algorithm for solving Bayesian Stackelberg games," in Proceedings of the 7th International Joint Conference on Autonomous Agents and Multiagent Systems, vol. 2, 2008, pp. 895-902.

[10] C. Strasburg, N. Stakhanova, S. Basu, and J. Wong, The Methodology for Evaluating Response Cost for Intrusion Response Systems, Technical Report 08-12, Iowa State University, 2008.

[11] Y. Zhang, L. Wang, W. Sun, R. C. Green, and M. Alam, "Distributed intrusion detection system in a multi-layer network architecture of smart grids," IEEE Transactions on Smart Grid, vol. 2(4), pp. 796-808, 2011.

[12] S. A. Zonouz, H. Khurana, W. H. Sanders, and T. M. Yardley, "RRE: A game-theoretic intrusion response and recovery engine," in Proc. IEEE/IFIP International Conference on Dependable Systems & Networks (DSN'09), 2009, pp. 439-448.

[13] C. A. Carver, Intrusion Response Systems: A Survey, Department of Computer Science, Texas A & M University, College Station, TX, 2000.

Understanding the Role of Automated Response Actions to Improve AMI Resiliency

Ahmed Fawaz[1], Robin Berthier[1], Bill Sanders[1], Partha Pal[2]
April 24, 2012

[1]University of Illinois at Urbana Champaign
[2]BBN Technologies

 TCIPG.ORG | 1 TRUSTWORTHY CYBER INFRASTRUCTURE FOR THE POWER GRID

TCIPG Mission

- Identify and address critical security and resiliency needs at the cyber-physical junction in the evolving power grid
  - Meet the challenge of rapid evolution and mixed legacy environment
  - Address the proliferation of devices, demand response, DG integration, HAN...
  - Emphasis on trust and resiliency
- Engage Industry (utility, control system vendors, technology providers)
  - Ensure relevance of research
  - Foster technology transfer

- Research Excellence
  - Balance long-range basic research with the need to develop practical solutions in the near term
  - Publications and conference presentations
  - TCIPG is the "go to" academic center
- Education
  - Develop university students who will be experts in the field
  - Outreach to K-12 students and the public

 TCIPG.ORG | 2 TRUSTWORTHY CYBER INFRASTRUCTURE FOR THE POWER GRID

TCIPG Statistics

- Builds upon $7.5M NSF TCIP CyberTrust Center 2005-2010
- $18.8M over 5 years, starting Oct 1, 2009
- Funded by Department of Energy, Office of Electricity and Department of Homeland Security
- 5 Universities
  - University of Illinois at Urbana-Champaign
  - Washington State University
  - University of California at Davis
  - Dartmouth College
  - Cornell University
- 20 Faculty, 20 Senior Technical Staff, 37 Graduate Students, 5 Undergraduate Students, and 1 Admin

 TCIPG.ORG | 3   TRUSTWORTHY CYBER INFRASTRUCTURE FOR THE POWER GRID

Industry Interaction: Vendors and Utilities that have participated in TCIPG Events

TCIPG.ORG | 4   TRUSTWORTHY CYBER INFRASTRUCTURE FOR THE POWER GRID

Agenda

- Background on AMI
  - System overview
  - Security aspects
- Towards Automated Response
  - Taxonomy
  - Cost model
  - Practical deployment
- Future Directions

 TCIPG.ORG | 7   TRUSTWORTHY CYBER INFRASTRUCTURE FOR THE POWER GRID

Acknowledgements

- Core funding as part of TCIPG Center, Office of Electricity, Department of Energy
- Additional support for Development/Analysis/ Technology Transfer
  - BBN Technology
  - ITRON – Testbed provisioning

 TCIPG.ORG | 8   TRUSTWORTHY CYBER INFRASTRUCTURE FOR THE POWER GRID

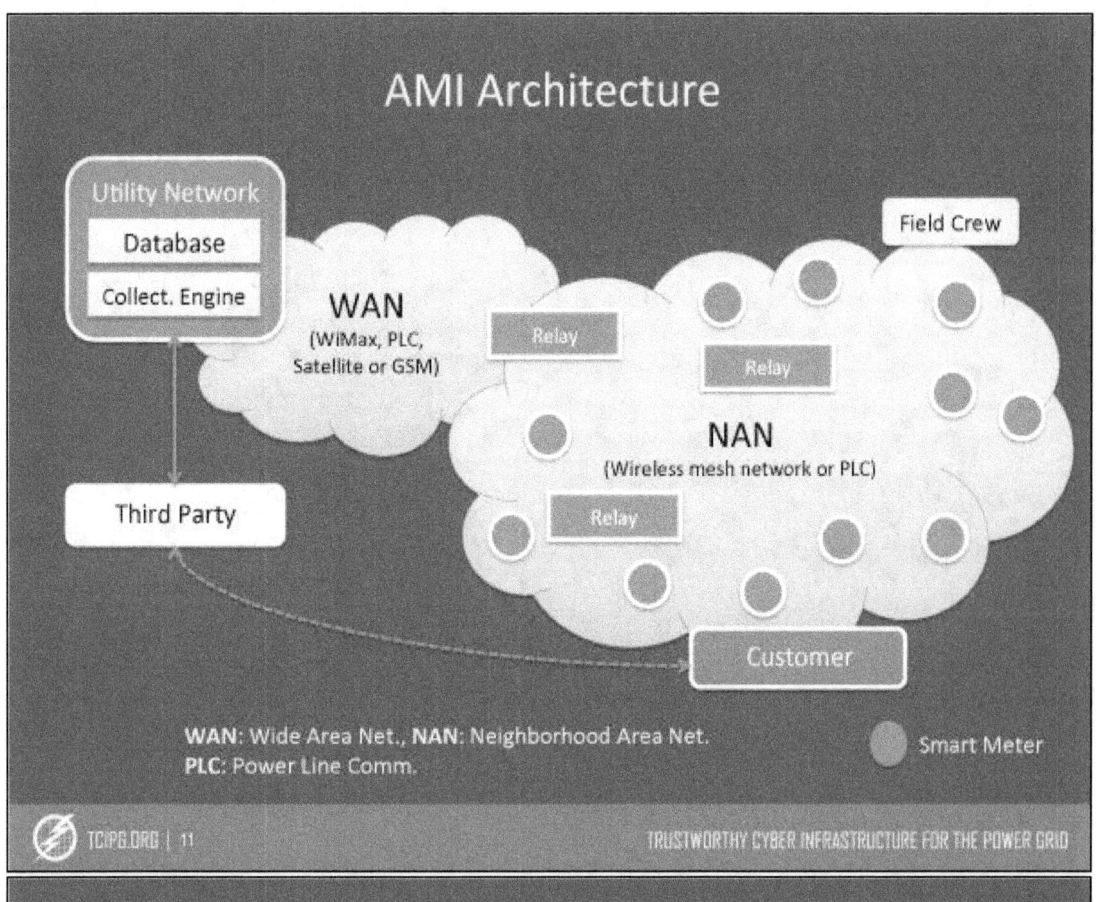

Requirements

* Large-scale
  - Managing few millions nodes

* Resilient
  - Energy delivery mission is critical

* Privacy-preserving
  - Protect sensitive customer information

TCIPG.ORG | 12   TRUSTWORTHY CYBER INFRASTRUCTURE FOR THE POWER GRID

Constraints

- Long term deployment
  - Life cycle of 5 to 15 years (vs. 2-3 years in IT)
- Meters have low-computational power
- Limited network bandwidth
- Limited information about attacks
- Security solutions should be:
  - Non-intrusive
  - Low maintenance

 TCIPG.ORG | 13   TRUSTWORTHY CYBER INFRASTRUCTURE FOR THE POWER GRID

Cyber Security Threats

- Motivations:
  - Energy fraud
  - Denial of service
    - Extortion
  - Power Disruption
    - Targeted remote disconnect
    - Large-scale outages and instability
  - Stealing personal information
  - Abuse of communication infrastructure
  - Loss of customer trust and adoption

 TCIPG.ORG | 14   TRUSTWORTHY CYBER INFRASTRUCTURE FOR THE POWER GRID

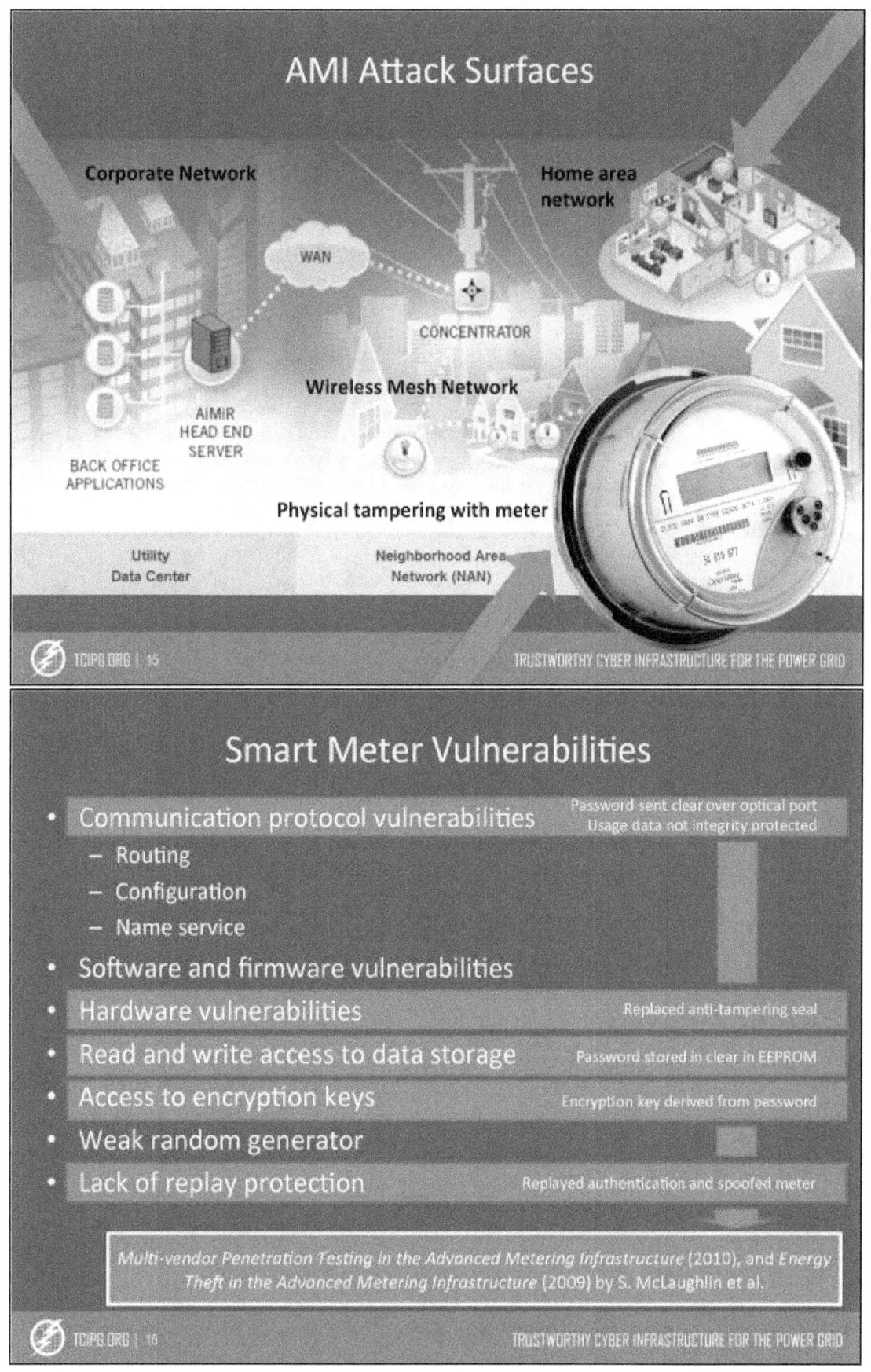

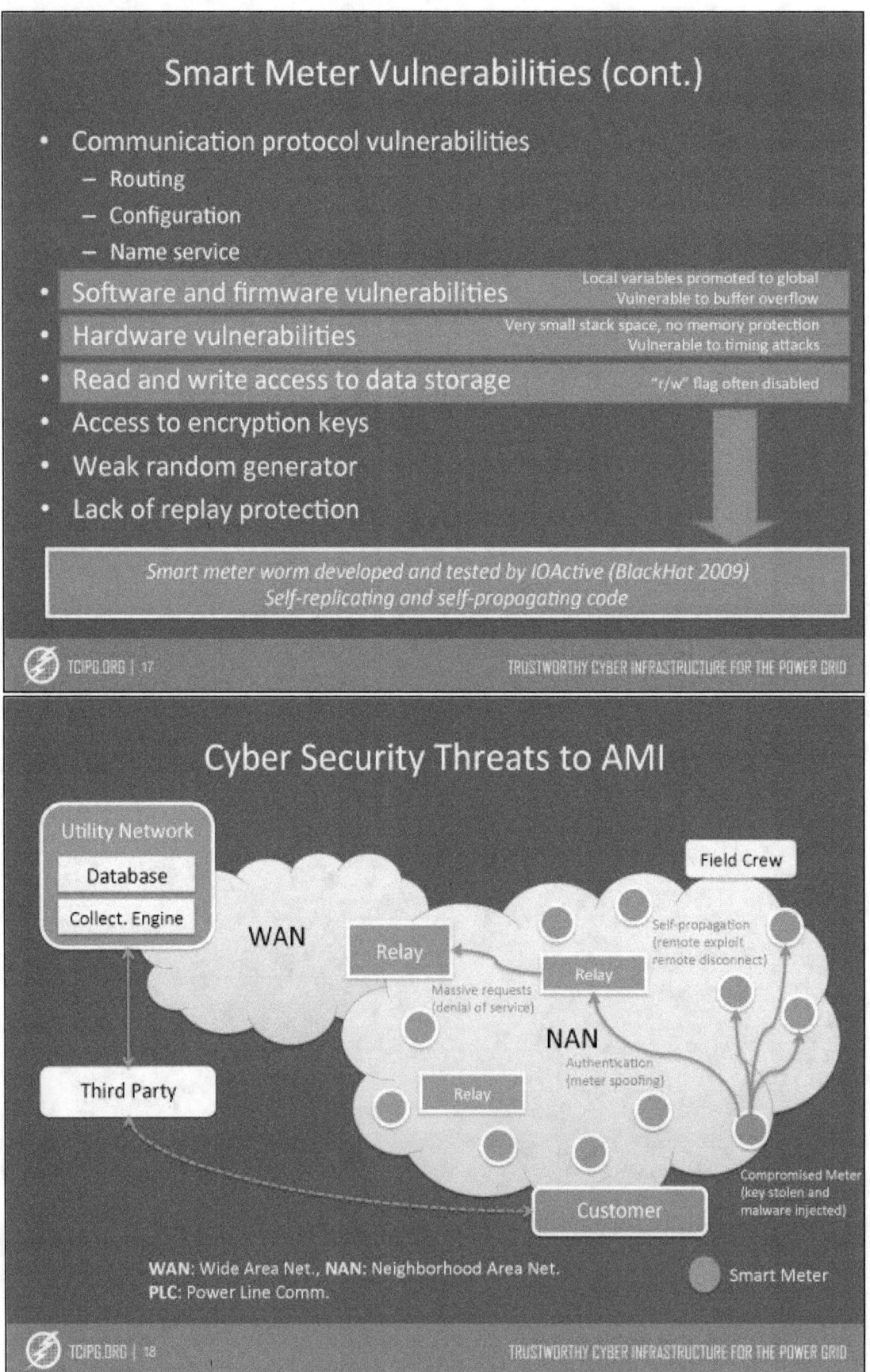

Multi-layered Security Approach

- Prevention
  - Authentication
  - Encryption
- Detection
  - Meter alarms/logs
  - Intrusion detection
- Response
  - Access control lists
  - Credentials/keys update
  - Firmware update

*Building a **resilient** architecture requires to implement all three*

 TCIPG.ORG | 19   TRUSTWORTHY CYBER INFRASTRUCTURE FOR THE POWER GRID

Multi-layered Security Approach

- Prevention
  - Authentication
  - Encryption
- Detection
  - Meter alarms/logs
  - Intrusion detection
- Response
  - Access control lists
  - Credentials/keys update
  - Firmware update

 TCIPG.ORG | 20   TRUSTWORTHY CYBER INFRASTRUCTURE FOR THE POWER GRID

Multi-layered Security Approach

- Prevention
  - Authentication
  - Encryption
- Detection
  - Meter alarms/logs
  - Intrusion detection
- **Response**
  - Access control lists
  - Credentials/keys update
  - Firmware update

Critical need for smart automated response
- *Complexity of large-scale distributed systems*
- *Efficiency of automated attacks*
- *Reduction of response cost and time*

 TCIPG.ORG | 21   TRUSTWORTHY CYBER INFRASTRUCTURE FOR THE POWER GRID

Work Plan

1. Understand possible response actions
   → Identify a taxonomy of AMI-specific actions

2. Understand safety/cost/benefit tradeoffs of actions
   → Define a cost model

3. Test and study practical deployment
   → Implement automated responses in TCIPG testbed

 TCIPG.ORG | 22   TRUSTWORTHY CYBER INFRASTRUCTURE FOR THE POWER GRID

RESPONSE ACTION TAXONOMY

   TRUSTWORTHY CYBER INFRASTRUCTURE FOR THE POWER GRID

Response Action Taxonomy

- Comprehensive response classification
  - Ensures coverage and completeness

- Customized for AMI
  - Cooperative actions among meters
  - Tunable response intensity
  - Special AMI recovery actions

- Understand characteristics of actions
  - Important for cost computation

   TRUSTWORTHY CYBER INFRASTRUCTURE FOR THE POWER GRID

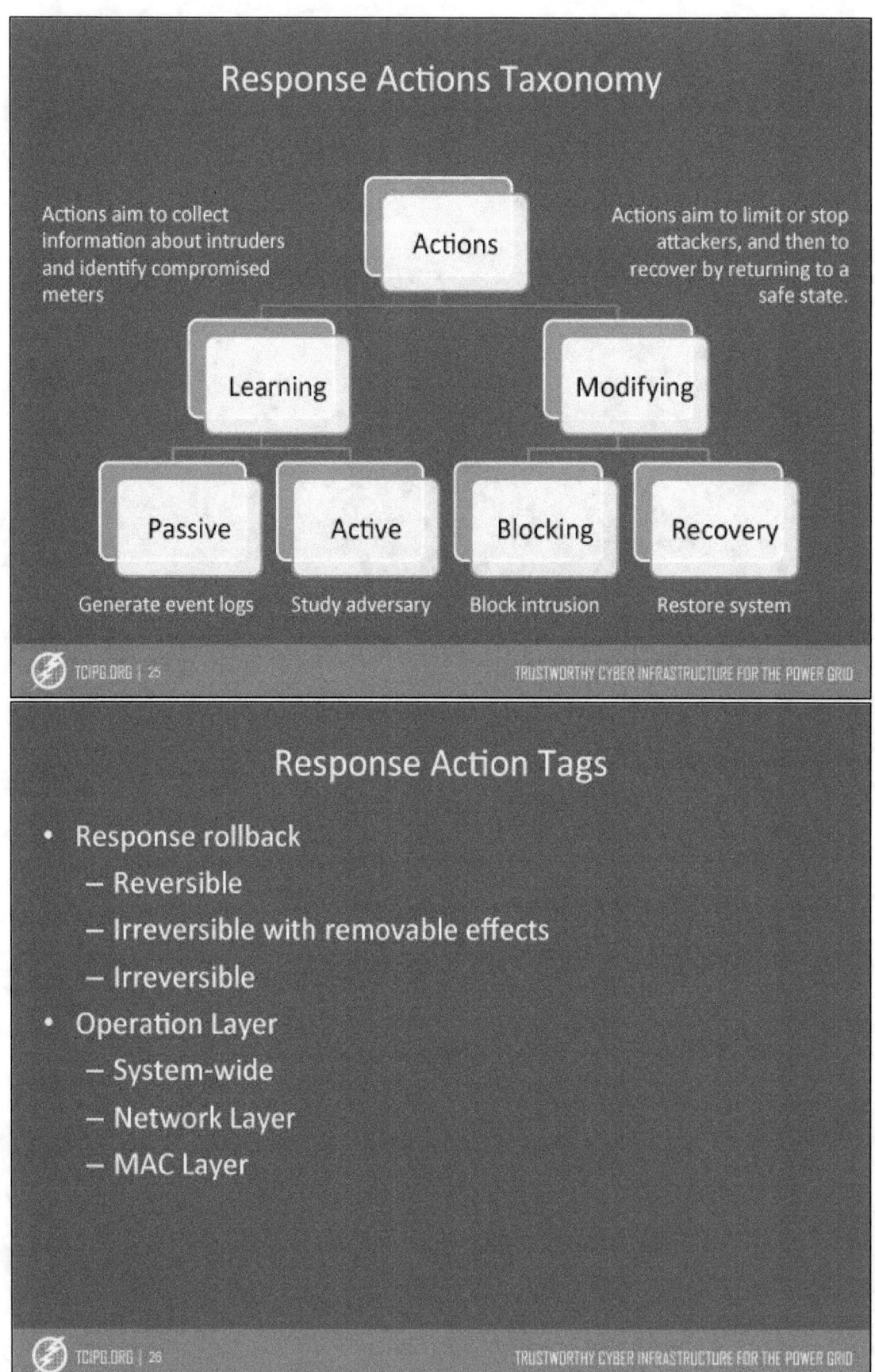

Response Action Tags

- Resources Involvement
  - Multiple meters
  - Single meter
  - Cooperative
- Admin Involvement
  - Fully automated
  - Requires admin input
- Response flexibility

 TCIPG.ORG | 27   TRUSTWORTHY CYBER INFRASTRUCTURE FOR THE POWER GRID

(Subset of the) Taxonomy

		Action	Rollback	Layer	Resources	Admin Involvement
Learning	Passive	Generate reports	N/A	System	Multiple	Automated
		Alarm	N/A	System		Automated
		Profile customers' power usage to detect anomalies	N/A	System	Multiple	Automated
	Active	Start analysis tools	N/A	Network	Multiple	Automated
		Verify ARP table entries (MAC-device mappings)	N/A	Network	Multiple	Automated
		Detect duplicates by probing the network	N/A	Network	Cooperative	Automated
		Send probe packets to test routes	N/A	Network	Cooperative	Automated
		Add decoy nodes	R	System		Automated
Modifying	Limiting	Isolate neighborhood	R	System	Multiple	Semi-automated
		Firewall rule at collector	R	Network	Single	Automated
		Blocking connections	IR	Network	Single	Automated
		Limiting network access	R	Network	Single	Automated
		Rate limiting network traffic	R	Network	Single	Automated
		Enabling quarantine / jail environment	R	System	Multiple	Automated
	Recovery	Merge neighborhood network temporarily	R	System	Multiple	Semi-automated
		Distribute attack signature	IR	System	Cooperative	Automated
		Verify C12.22 routing tables	N/A	Network	Cooperative	Automated
		Apply patch	IR	System	Multiple	Semi-automated
		Replace meter (physically)	IR	N/A	Single	Semi-automated
		Recover meter readings	N/A	N/A	Multiple	Automated
		Turn on/off service (recover attack)	R	N/A	Multiple	Automated

 TCIPG.ORG | 28   TRUSTWORTHY CYBER INFRASTRUCTURE FOR THE POWER GRID

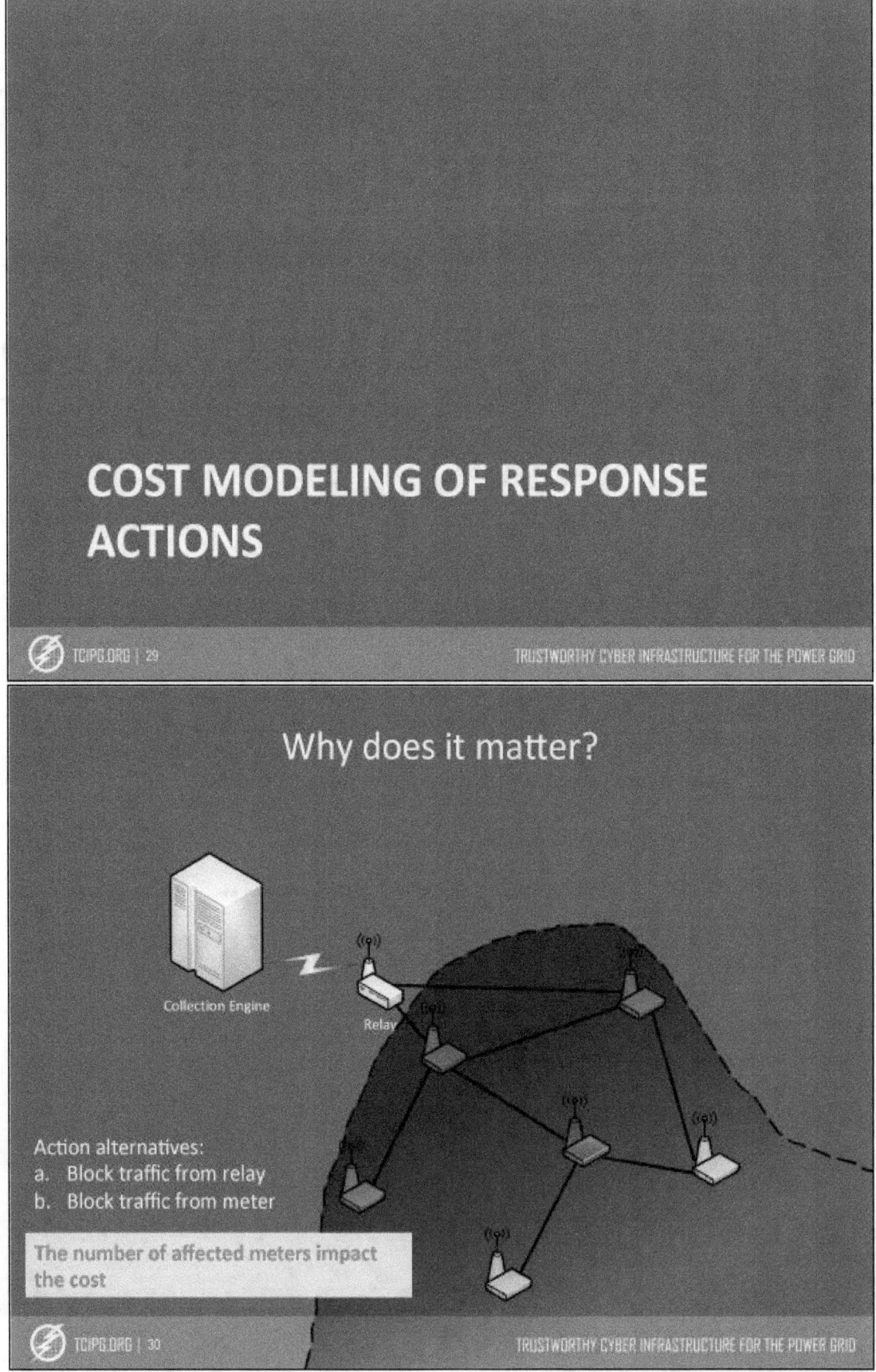

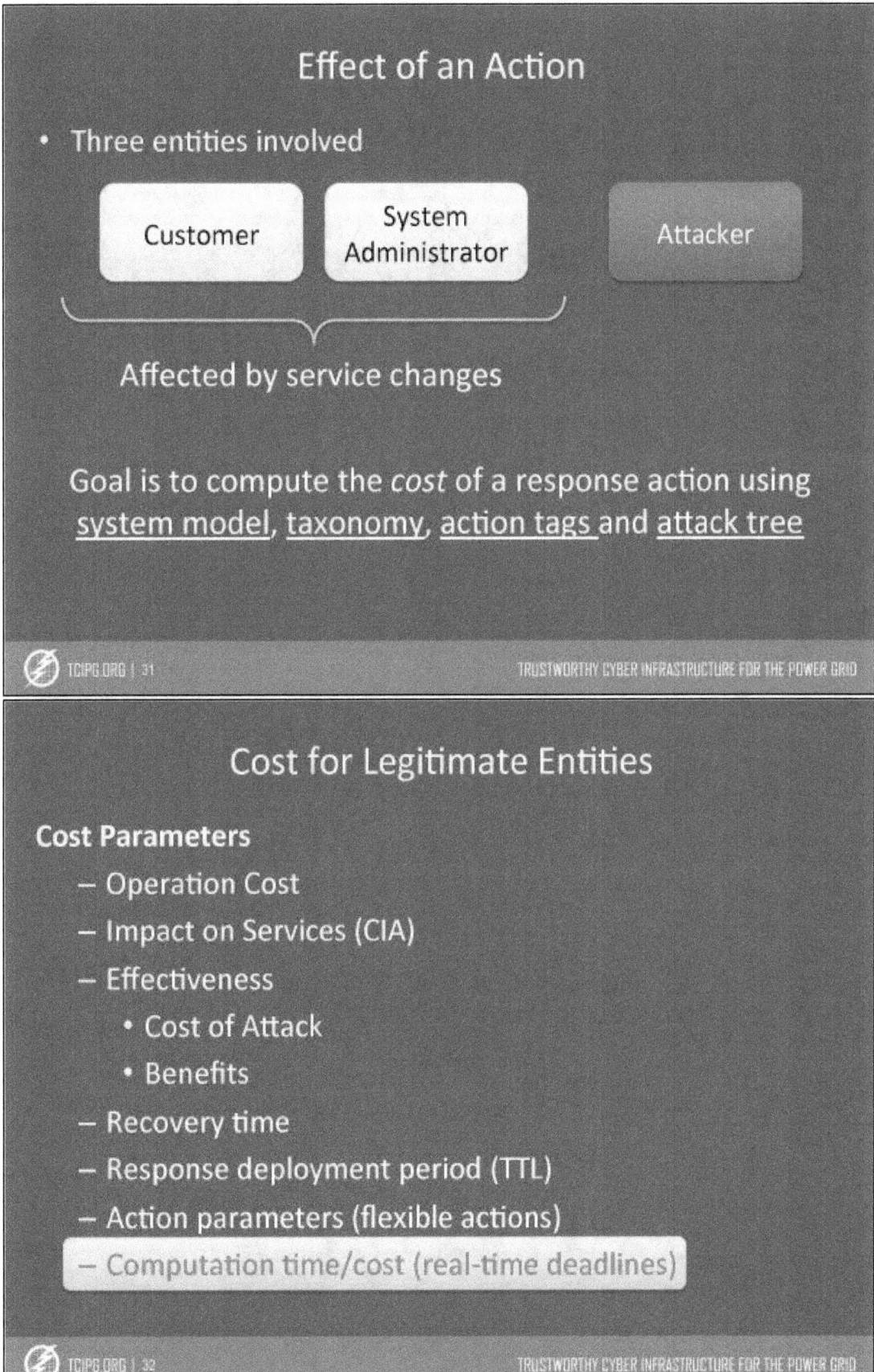

Effect of an Action

- Three entities involved

| Customer | System Administrator | Attacker |

Affected by service changes

Goal is to compute the *cost* of a response action using
<u>system model</u>, <u>taxonomy</u>, <u>action tags</u> and <u>attack tree</u>

Cost for Legitimate Entities

Cost Parameters
- Operation Cost
- Impact on Services (CIA)
- Effectiveness
  - Cost of Attack
  - Benefits
- Recovery time
- Response deployment period (TTL)
- Action parameters (flexible actions)
- Computation time/cost (real-time deadlines)

Current Approaches

Current approaches capture a partial image

- Static costs mapped to actions
    - Systems dynamics alter the effect of an action

- Parameterized cost
    - Operation cost, damage cost, response goodness and impact (static parameters)
    - Ensures better coverage but does not capture system dynamics

- Resource dependency model
    - Capture dynamics but leads to an incomplete cost value

 TCIPG.ORG | 33   TRUSTWORTHY CYBER INFRASTRUCTURE FOR THE POWER GRID

Computing Effect on CIA

- System modeled using *Dependency graph G=(V,E)*
    - V set of resources
    - E set of edges (*r, s*) representing relation

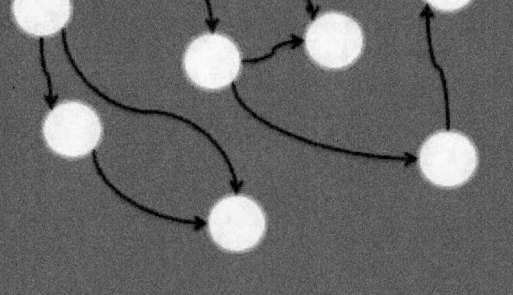

- Resources labeled with dysfunction rate vector

    Vr[C,I,A]

- Each edge labeled with a degree matrix $W_{r_i r_j} = (\blacksquare w_{r_i r_j}(1,1) \ \& w_{r_i r_j}(1,2) \ \& w_{r_i r_j}$

 TCIPG.ORG | 34   TRUSTWORTHY CYBER INFRASTRUCTURE FOR THE POWER GRID

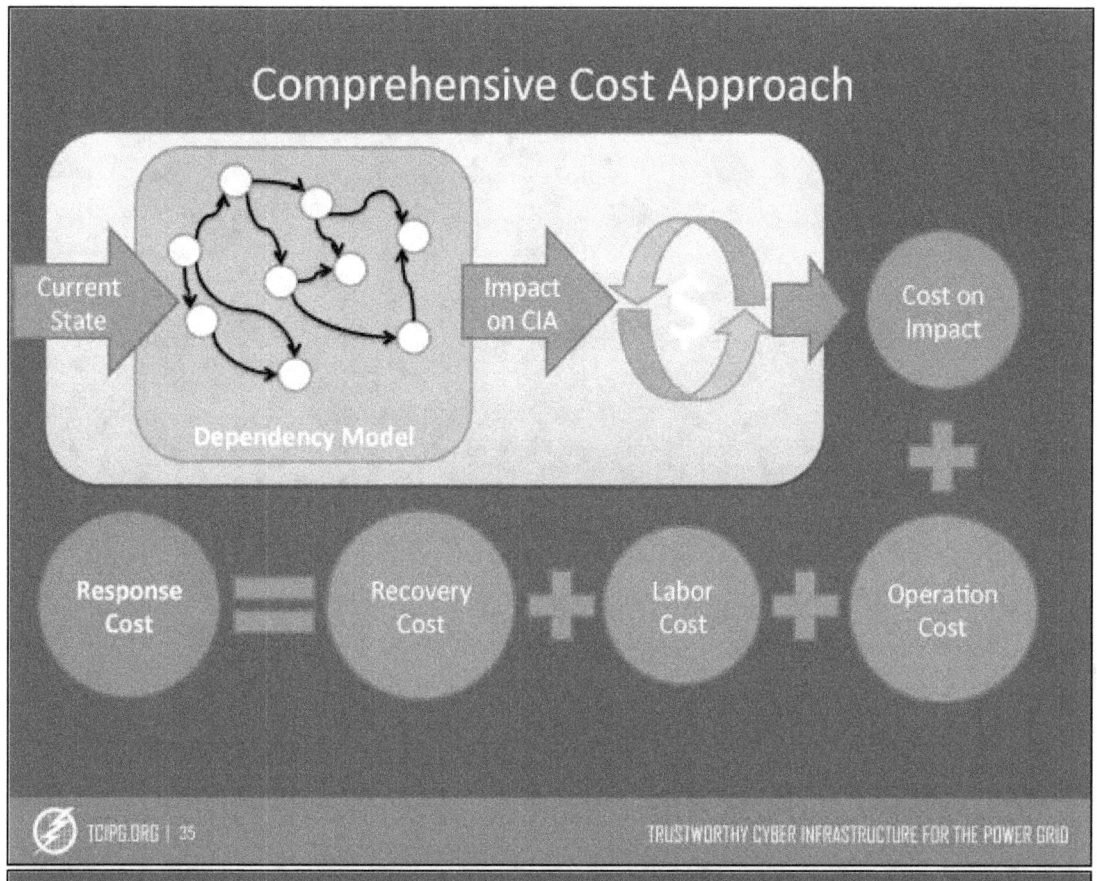

Comprehensive Cost Approach

(Slide: Current State → Dependency Model → Impact on CIA → $ → Cost on Impact)

Response Cost = Recovery Cost + Labor Cost + Operation Cost

Cost Effectiveness

Taxonomy characterizes two high level type
- *Learning*: leaves attack running
- *Modifying*: activity tackles attack

Impact of attack on system
- Model attacker (Möbius ADVISE model)
  - Objectively simulates multiple adversary models
- Probabilistic attack costs
  - Tag attack trees with costs
- Historic data
  - Not enough for complete cost model

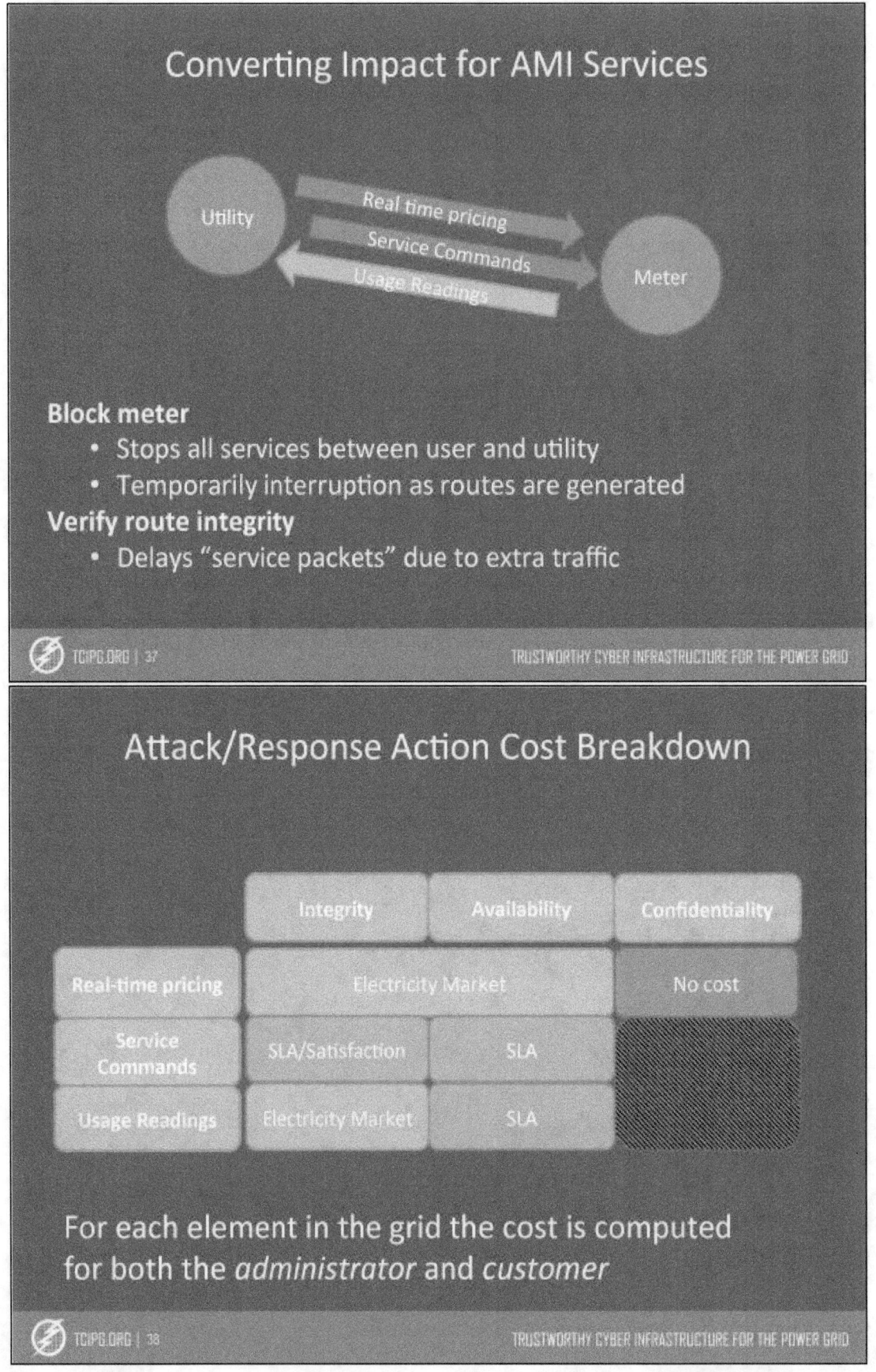

Availability for Pricing Information

- Customer uses flat rate in case of unavailable pricing information

- *Flat rate< Market rate*
  - Utility loses revenue
- *Flat rate> Market rate*
  - Customer overcharged

$$Cost=\Delta(price)\times Usage$$

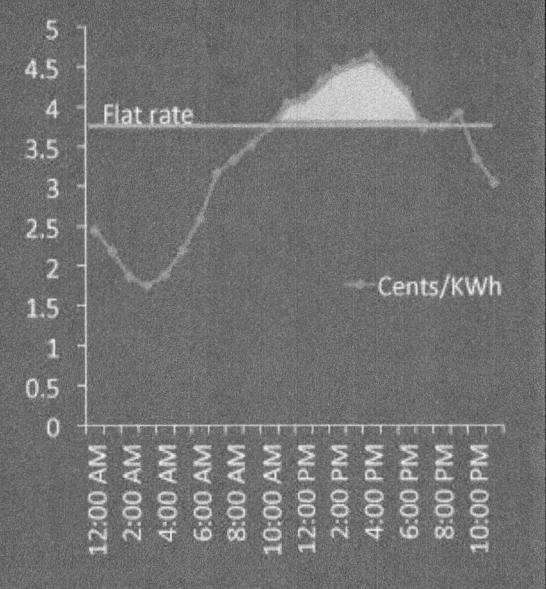

TCIPG.ORG | 39   TRUSTWORTHY CYBER INFRASTRUCTURE FOR THE POWER GRID

Integrity of Pricing Information

- Action increases rate
  - Customer dissatisfaction

- Action lowers rate
  - Customer over billed (legal action)
  - Increase demand for power
    - Rate increase
    - Generation perturbations

TCIPG.ORG | 40   TRUSTWORTHY CYBER INFRASTRUCTURE FOR THE POWER GRID

Impact on Meter Readings

- Availability
  - SLA penalty
  - Delay in EMS usage profiles

- Integrity
  - Energy theft or overbilling customer
  - Misleading usage profiles

 TCIPG.ORG | 41   TRUSTWORTHY CYBER INFRASTRUCTURE FOR THE POWER GRID

Impact on Service Commands

- Availability
  - SLA penalty
  - Customer dissatisfaction due to delays in utility services (turning on power, blackouts detection,...)

- Integrity
  - Extra labor and operation costs due to false positives
  - Cost increase for the customer

 TCIPG.ORG | 42   TRUSTWORTHY CYBER INFRASTRUCTURE FOR THE POWER GRID

Cost of Confidentiality

- Compromised confidentiality
  - Leads to invasion to privacy through load profiling

- Legal action and lost confidence

- Current surveyed SLAs *do not* contain provisions for confidentiality

 TCIPG.ORG | 43   TRUSTWORTHY CYBER INFRASTRUCTURE FOR THE POWER GRID

Provisions for Service Level Agreements

- Availability

  Guarantee that usage data, commands and pricing arrive in a timely manner within regular load

- Integrity

  Guarantee that X% of usage data, commands and pricing are not tampered

- Confidentiality

  Guarantee that X% of usage data privacy is not compromised

 TCIPG.ORG | 44   TRUSTWORTHY CYBER INFRASTRUCTURE FOR THE POWER GRID

Cost for Attacker

- Stop attack
  - Block compromised entities

- Slow down attack
  - Rate limiting of compromised entities

- Facilitate attack
  - Misdiagnosis or misconfigured response
  - Collect information on the attacker and the strategies used

 TCIPG.ORG | 45   TRUSTWORTHY CYBER INFRASTRUCTURE FOR THE POWER GRID

IMPLEMENTATION & DEPLOYMENT

 TCIPG.ORG | 46   TRUSTWORTHY CYBER INFRASTRUCTURE FOR THE POWER GRID

Framework

- Intrusion response systems can be based on:
  - Heuristics
  - Machine learning
  - Game theory

Recovery and Response Engine (RRE)

- Assumes security game between attacker and defender
- Uses an Attack-Response tree to model system
- Computes the optimal response strategy that minimizes the cost for an administrator
- $r(s,a,s')=(\delta{\downarrow}g\,(s)-\delta{\downarrow}g\,(s'))\hat{} \tau{\downarrow}1\ \ C(a)\hat{} \tau{\downarrow}2$
  - **C(a)** is the cost function introduced by our cost model

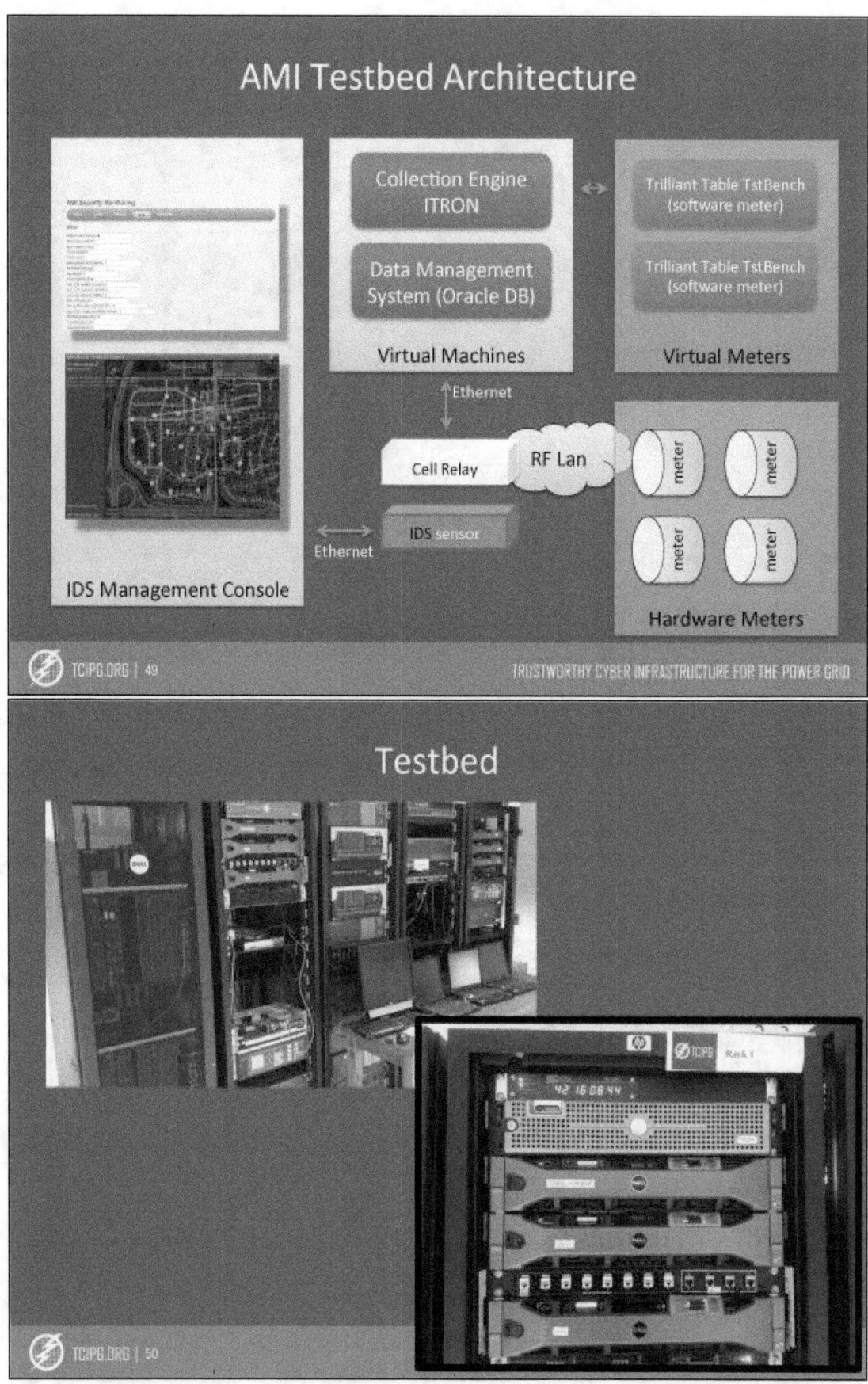

Future Work

- Automate the assignment of weights in dependency model using minimum administrator input
- Automate the generation of relations between CIA
- Complete case study by defining models for the different security implications
- Design a "security inspired" metering SLA
- Complete implementation in RRE framework
- Initiate testing within testbed using realistic AMI
- Optimize performance

 TCIPG.ORG | 53   TRUSTWORTHY CYBER INFRASTRUCTURE FOR THE POWER GRID

Safety

- Attackers can drive automated responses by triggering IDS sensors

- A separate unit to include the admin to the loop in the case for some specified actions

- Actions with safety issues should be semi-automated
  - Provide a choice for the admin with alternatives

- Define a safety criterion for AMI

 TCIPG.ORG | 54   TRUSTWORTHY CYBER INFRASTRUCTURE FOR THE POWER GRID

Research Questions

- Design modular response actions and cost model
  - Ensures compatibility with different technologies and implementations

- Automate generation of response actions

- Propose Performance metrics for automated responses

 TCIPG.ORG | 55   TRUSTWORTHY CYBER INFRASTRUCTURE FOR THE POWER GRID

Conclusion

- Formed a response action taxonomy with learning and modifying categories
- Current cost models rely on subjective administrator parameters or static values
- Defined response cost model to include parameters from the taxonomy
- Map response parameters to monetary values using SLAs and other cost factors
- Plan to implement automated response for AMI testbed

 TCIPG.ORG | 56   TRUSTWORTHY CYBER INFRASTRUCTURE FOR THE POWER GRID

Questions?

Robin Berthier <u>rgb@illinois.edu</u>

Ahmed Fawaz <u>afawaz2@illinois.edu</u>

William H. Sanders <u>whs@illinois.edu</u>

TCIPG.ORG | 57   TRUSTWORTHY CYBER INFRASTRUCTURE FOR THE POWER GRID

12. Cyber-Physical Systems Security for the Smart Grid

Alvaro A. Cárdenas
Fujitsu Laboratories of America
Email: alvaro.cardenas-mora@us.fujitsu.com

Ricardo Moreno
Universidad de los Andes, Colombia
Email: r-moreno@uniandes.edu.co

I. INTRODUCTION

The integration of Information Technology (IT) systems (computations and communications–the cyber world) with sensor and actuation data (the physical world), can introduce new, and fundamentally different approaches to security research in the growing field of Cyber-Physical Systems (CPS), when compared to other purely-cyber systems. In our earlier work [, , ,], we have shown that because of the automation and real-time requirements of many control actions, traditional security mechanisms are not enough for protecting CPS, and we require *resilient control and estimation* algorithms for true CPS defense-in-depth.In this abstract we outline how attacks and resilient mechanisms can affect and defend power grid operations.

II. ENERGY MANAGEMENT SYSTEMS

To characterize the CPS security of the power grid, we need to understand how IT is used in the control centers of the power grid to collect sensor data, estimate the state of the power grid, and issue control commands and pricing signals to the market.

One of the most important components in a control center is the Energy Management System (EMS). The EMS is responsible for many operational tasks. It includes the Network Topology Processor (NTP), state estimation, the market process that delivers Locational Marginal Pricing (LMP), and control actions for transmission automation, such as remote tap adjustment for transformers.

The role of NTP, and state estimation is to collect data from sensors in the field, and give an accurate view (topology and electricity flows). If the data collected is incorrect, operators will get an erroneous view of the system and all management functions of the control center will be affected, (including market computations and control actions). This is the reason why a lot of recent work has focused on deception attacks (also known as false-data injection), where a compromised sensor sends malicious data back to the EMS).

In this abstract we survey recent work on CPS security for power systems and present the work in a unified view by showing how all previous attacks are part of the EMS. We find some limitations with previous work and discuss open problems and new research challenges of parts of the EMS that have not been considered in previous work.

A. State Estimation

The state estimation problem in power systems originates from the need of power engineers to estimate the phase angles $x \in \mathbb{R}^n$ from the measured power flow $z \in \mathbb{R}^m$ in the transmission grid. It is known that the measured power flow $z = h(x) + e$ is a nonlinear noisy measurement of the state of the system x and an unknown quantity e called the measurement error. To estimate x from this set of equations engineers make usually two simplifications: (1) e is assumed to be a Gaussian noise vector with zero mean and covariance matrix W, and (2) the equation is approximated by the linear equation

$$z = Hx + e$$

(where H is a matrix). Estimating x from these equations is achieved by computing the Minimum Mean Square Error (MMSE) estimate:

$$\hat{x} = (H^T W H)^{-1} H^T W z.$$

Because a typical transmission line system is composed of thousands of sensors (i.e., z is a vector of thousands of scalars) and not all sensors are reliable, power engineers have devised a set of tests to detect bad measurements. The tests are based on the following test:

$$||z - H\hat{x}|| > \tau \qquad (1)$$

that is, if the measurement and the estimated measurement are greater than a threshold, then the test decides that there are some faulty sensors in the transmission line. If the test has a value lower or equal to τ then the test concludes that all measurements are correct.

Liu et.al. [ ] introduced attacks against the integrity of state estimation algorithms in the power grid by showing that there are attacks where a compromised sensor can send a false measurement reading and yet the bad data detection test will not detect this attack. In particular, they show how by selecting an attack signal $z_a = z + a$, where $a = Hc$ (for any vector c) creates a successful attack. Then they analyzed how attackers can craft these attacks when they have different resources (limited access to meters or limited ability to compromise meters) and different objectives (random attacks or specific errors in the estimate). While attacks in larger systems are difficult to create (in an IEEE 3000 bus system the attacker needs to compromise more than 900 meters) and may have limited negative effects (the injected error might not be too large), the fact that attackers can manipulate the view of one of our critical infrastructures is a worrisome fact.

Some follow up work has discussed extensions on how to better protect the power grid to these attacks.

Some preliminary results in this area of research include the work of Dán and Sandberg [ ], who consider a defender that can secure individual measurements by, for example, replacing an existing meter to a meter with better security mechanisms such as tamper resistance or hardware security support. Their goal is to protect the system under a limited budged and to that end they formulate the problem as identifying the best k measurements to protect (they assume the attacker cannot compromise these sensors) in order to minimize the impact of attacks. The mathematical problem they consider is a combinatorial optimization, so this problem is intractable for large systems. The main contribution of this work is to exploit the structure of the power system matrices to make the optimization problem efficient. Kosut et.al. [ ] also extend the basic false data injection attack to consider attackers trying to maximize the error introduced in the estimate, and defenders with a new detection algorithm that attempts to detect false data injection attacks. Their new detection algorithm performs better than the traditional bad data detection algorithms (since these algorithms were designed for detecting faults, not network attacks). Their detection algorithm is based on the generalized likelihood ration test, which is not a tractable problem to solve.

B. Network Topology Processor

Each breaker in the transmission system has a sensor reporting if it is open or closed. This information is sent to the NTP to construct the topological model of the system. This topological model is used for the state estimation of the system. If the topology is wrong, the state estimation algorithm will also produce erroneous results.

As far as we know, no previous work has studied the false-data injection problem against the NTP.

C. Electricity Markets

The goal of the electricity market process in the control center is to deliver LMPs. LMPs are computed at each load and at each generation point when the transmission system is congested (which is the default state) to determine how much will utilities pay the system operator (per Megawatt), and how much will the system operator pay the generation points. LMPs are traditionally computed every 5 to 10 minutes, but there is recent work (e.g., New York power system) for computing LMPs in real-time.

Quantifying the cost of security incidents is one of the most difficult problems in computer security because it is hard to quantify the value of information. However, by analyzing attacks against the electricity market, we can quantify the effects of these attacks by leveraging the economic metrics used to measure the efficiency of the system.

As we mentioned before, if the state estimation is incorrect, all management functions of the control center are affected, including the market operations. Xie et.al. [ ] studied how false data injection attacks can be used to defraud deregulated electricity markets by modifying LMPs. They consider the case where attackers can manipulate prices while being undetected by the system operator.

In all other attacks considered in this paper, the attacker can be implicitly assumed to be a malicious entity that tries to destabilize the system or reduce the social welfare. on the other hand, the work of Xie et.al., considers a *selfish attacker* instead of a *malicious attacker*. This change in the motivation of the attacker makes it difficult to understand which party will have the long-term motivation to launch these type of attacks. Utilities, generators, and system operators are large, highly regulated companies who have higher incentives to remain in business than to launch an attack that can put their company in jeopardy (in case it is discovered).

So far, all attacks presented in this abstract were based on false-data injection attacks against the sensor data used for state estimation. Negrete-Pincetic et.al [ ] consider a new type of attacks by studying the integrity of the control signals (as opposed to the integrity of the sensor signals). In particular, they study how malicious control signals sent to circuit breakers (directing them to remove transmission lines from the system) affect the social welfare metric of the market system.

D. Transmission Automation

In addition to the control signals sent to circuit breakers, as considered by Negrete-Pincetic et.al., there are many other control signals that can be falsified by an attacker, in particular, given that the smart grid is introducing the capability of more distributed, automatic control.

The Flexible Alternate Current Transmission System (FACTS) includes many automatic electronic devices such as Static Voltage Compensators (SVC), which similar to capacitor banks, uses reactive power to improve the voltage profile of the system. Similarly, the Thyristor Controlled Series Compensator (TCSC) is a control devices in series with a transmission line which can be used to modify its impedance to control the current going through these lines. A taxonomy of attacks against FACTS devices was presented by Phillips et.al. [ ] and an implementation of some attacks with false status reports and control actions showed unnecessary VAR compensation and unstable operation of the sytem [ ]

Other control signal that can be sent remotely include tap adjustments for smart transformers (used to increase or decrease slightly the voltage on each side of a transformer), and the Automatic Generator Control (AGC) signal (which is used to set the voltage of generators). Robust attack policies have been studied for AGC signals [,], and attacks have shown that if you modify the frequency and tie-line flow measurements, the system can be driven to abnormal operating values [ ].

III. Defense Mechanisms

In addition to traditional IT security mechanisms for prevention (authentication, encryption, firewalls) and detection (intrusion detection systems, forensics) we need new CPS security mechanisms.

There are several CPS planning and defense mechanisms that can leverage knowledge of the attacks presented in this abstract. The first is *risk assessment*: given a fixed budget, where should I allocate this budget to minimize my potential *physical* damages?

A second mechanism is *bad data detection* mechanisms. These mechanisms should not assume random, independent failures, but consider detection of sophisticated attackers. Interestingly enough, most previous work has focused on attacks and the quantification of these attacks, but very few have proposed novel attack-detection mechanisms [ ]. One particular open problem is to propose *bad topology*

detection mechanisms.

Replacing sensed data with false data (a deception attack) is a very generic attack that can be extended to any smart grid application (as all of them are based on correct sensor measurements). It is important to develop intrusion detection mechanisms or reputation management systems for smart grid applications where not all received data can be trusted.

The defense third mechanism is to introduce *resiliency* (or survivability) of the system to attacks. A promising direction is to design the topology of the power distribution network to withstand malicious commands to circuit breakers trying to change and disconnect the network [ ].

IV. Conclusions

CPS security is a growing field critical for the vision of a survivable power grid that can withstand attacks and reconfigure or adapt to mitigate adverse effects.

Work on fault tolerance and reliability of control systems is not enough, because these mechanisms generally assume independent and uncorrelated failures; however, cyber-attacks will exploit vulnerabilities in a coordinated and correlated fashion. The most basic example is the work of of false-data injection attacks [ ], where it is shown that traditional safety and fault-detection mechanisms currently available in the power grid cannot detect incorrect sensor data when a malicious attacker is the source of these errors.

Therefore instead of relying solely on fault-detection algorithms to protect control algorithms in the power grid, we need to develop new attack-detection algorithms focusing on identifying malicious data in sensor and actuation devices in the power grid.

REFERENCES

[1] S. Amin, A. A. Cárdenas, and S. S. Sastry. Safe and secure networked control systems under denial-of-service attacks. In *HSCC '09: Proceedings of the 12th International Conference on Hybrid Systems: Computation and Control*, pages 31–45, Berlin, Heidelberg, 2009. Springer-Verlag.

[2] A. Cardenas, S. Amin, and S. Sastry. Research Challenges for the Security of Control Systems. In *3rd USENIX workshop on Hot Topics in Security (HotSec '08). Associated with the 17th USENIX Security Symposium.*, July 2008.

[3] A. A. Cárdenas, S. Amin, Z.-S. Lin, Y.-L. Huang, C.-Y. Huang, and S. Sastry. Attacks against process control systems: risk assessment, detection, and response. In *ASIACCS '11: Proceedings of the 6th ACM Symposium on Information, Computer and Communications Security*, pages 355–366, New York, NY, USA, 2011. ACM.

[4] A. A. Cardenas, S. Amin, and S. Sastry. Secure control: Towards survivable cyber-physical systems. In *Proceedings of the First International Workshop on Cyber-Physical Systems.*, June 2008.

[5] G. Dán and H. Sandberg. Stealth Attacks and Protection Schemes for State Estimators in Power Systems. In *First IEEE Smart Grid Commnunications Conference (SmartGridComm)*, October 2010.

[6] P. Esfahani, M. Vrakopoulou, K. Margellos, J. Lygeros, and G. Andersson. A robust policy for automatic generation control cyber attack in two area power network. In *Decision and Control (CDC), 2010 49th IEEE Conference on*, pages 5973 –5978, dec. 2010.

[7] O. Kosut, L. Jia, R. Thomas, and L. Tong. Malicious Data Attacks on Smart Grid State Estimation: Attack Strategies and Countermeasures. In *First IEEE Smart Grid Commnunications Conference (SmartGridComm)*, October 2010.

[8] Y. Liu, M. K. Reiter, and P. Ning. False data injection attacks against state estimation in electric power grids. In *CCS '09: Proceedings of the 16th ACM conference on Computer and communications security*, pages 21–32, New York, NY, USA, 2009. ACM.

[9] P. Mohajerin Esfahani, M. Vrakopoulou, K. Margellos, J. Lygeros, and G. Andersson. Cyber attack in a two-area power system: Impact identification using reachability. In *American Control Conference (ACC), 2010*, pages 962 –967, 30 2010-july 2 2010.

[10] R. Moreno and A. Torres. Network Topological Analysis to Assess the Power Smart Grid. In *2012 IEEE Innovative Smart Grid Technologies (ISGT) Conference*, January 2012.

[11] M. Negrete-Pincetic, F. Yoshida, and G. Gross. Towards quantifying the impacts of cyber attacks in the competitive electricity market environment. In *2009 IEEE PowerTech*, June 2009.

[12] L. Phillips, M. Baca, J. Hills, J. Margulies, B. Tejani, B. Richardson, and L. Weiland. Analysis of operations and cyber security policies for a system of cooperating flexible alternating current transmission system, Dec. 2005.

[13] S. Sridhar and G. Manimaran. Data integrity attacks and their impacts on SCADA control system. In *Prc. IEEE Power Energy Soc. General Meeting*, Jul. 2010.

[14] S. Sridhar and G. Manimaran. Data

193

integrity attack and its impacts on voltage control loop in power grid. In *Prc. IEEE Power Energy Soc. General Meeting*, Jul. 2011.

[15] L. Xie, Y. Mo, and B. Sinopoli. False Data Injection Attacks in Electricity Markets. In *First IEEE Smart Grid Commnunications Conference (Smart-GridComm)*, October 2010.

194

Securing Cyber-Physical Systems

Alvaro Cárdenas
Fujitsu Laboratories

Ricardo Moreno
Universidad de los Andes

From Sensor Nets to Cyber-Physical Systems

- Control
- Computation
- Communication

- Interdisciplinary Research!

- Example: Smart Grid

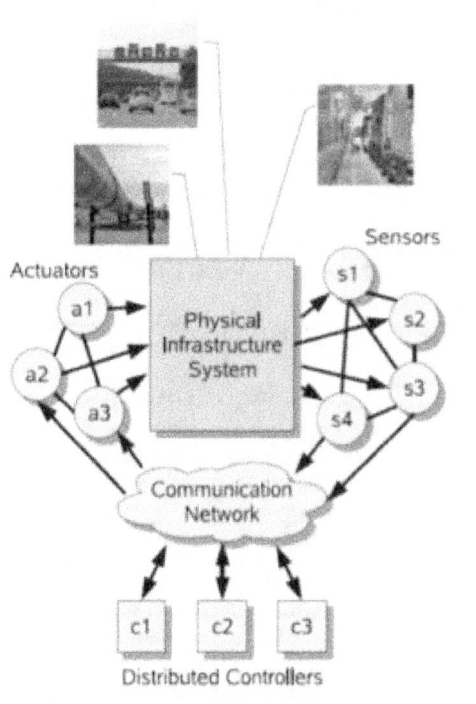

Attacks & Threats

■ Attacks

■ Maroochy Shire 00

■ HVAC 12

■ Stuxnet 10

■ Threats

Obama Adm Demonstrates In Feb. 2012 attack to power Grid

Securing CPS is Hard

■ Vulnerabilities are increasing

- ■ Sensors/Controllers are now computers (can be programmed for general purposes)
- ■ Networked (remotely accessible)
- ■ By necessity, billions of low-cost embedded devices
- ■ Physically insecure locations

■ Attacks will continue to happen

- ■ Devices deployed for ~ 20-30 years

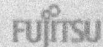

FBI: Smart Meter Hacks Likely to Spread

A series of hacks perpetrated against so-called "smart meter" installations over the past several years may have cost a single U.S. electric utility hundreds of millions of dollars annually, the FBI said in a cyber intelligence bulletin obtained by

Three Steps to Improve CPS Security

- ■ Short Term
  - ▧ Incentives
  - ▧ Software reliability
  - ▧ Solve basic vulnerabilities
- ■ Medium Term
  - ▧ Leverage Big Data for Situational Awareness
- ■ Long Term Research
  - ▧ Resilient estimation and control algorithms

Security is a Hard Business Case

- ■ *"Making a strong business case for cybersecurity investment is complicated by the difficulty of quantifying risk in an environment of rapidly changing, unpredictable threats with consequences that are hard to demonstrate"*
  - ▧ DoE Roadmap
- ■ Governments are responsible for Homeland Security, and critical infrastructure security
  - ▧ Utilities are not (outside their budget/scope?)
  - ▧ Problem:
    - · Interdependencies (e.g., cascading failures)
    - · It doesn't matter if one utility sets an example because this is a weakest security game
  - ▧ Nations have much more to lose from an attack than utilities

[Cardenas. CIP Report, GMU, 2012]

Short-term proposal

FUJITSU

- Vendors of equipment for managing control systems have few incentives for secure development programs because customers are not requesting them
- Asset owners need to request vendors secure coding practices, hardened systems, and quick response when new vulnerabilities and attack vectors are identified
- American Law Institute (ALI)
  - Principles of the Law of Software Contracts (2009)
  - Vendors liable for knowingly shipping buggy software
  - Implied warranty of no material hidden defects (non-disclaimable)
  - Software for CIP can be first use case
- Currently congress is debating how to give incentives for asset owners to invest in security
  - Cybersecurity Act 2012 (increase regulation)
  - SECURE-IT Act 2012 (increase data sharing)

Three Steps to Improve CPS Security

FUJITSU

- Short Term
  - Incentives
  - Software reliability
  - Solve basic vulnerabilities
- Medium Term
  - Leverage Big Data for Situational Awareness
- Long Term Research
  - Resilient estimation and control algorithms

Again, Security is a Hard Business Case

- **Push back in prices**
  - Billions of low-cost embedded devices
  - Can't have fancy tamper protection
- **Security is hard to see**
  - Hard to see advantages of hardening devices
- **But, Situational Awareness is Fun to see**
  - Understand the health of the system
    - Routing protocol, health of the system
  - Identify anomalies
- **Big Data is new in Smart Grid**
  - Redundancy
  - Diversity
  - Data Analytics to identify suspicious behavior

Big Data Analytics in Smart Grid

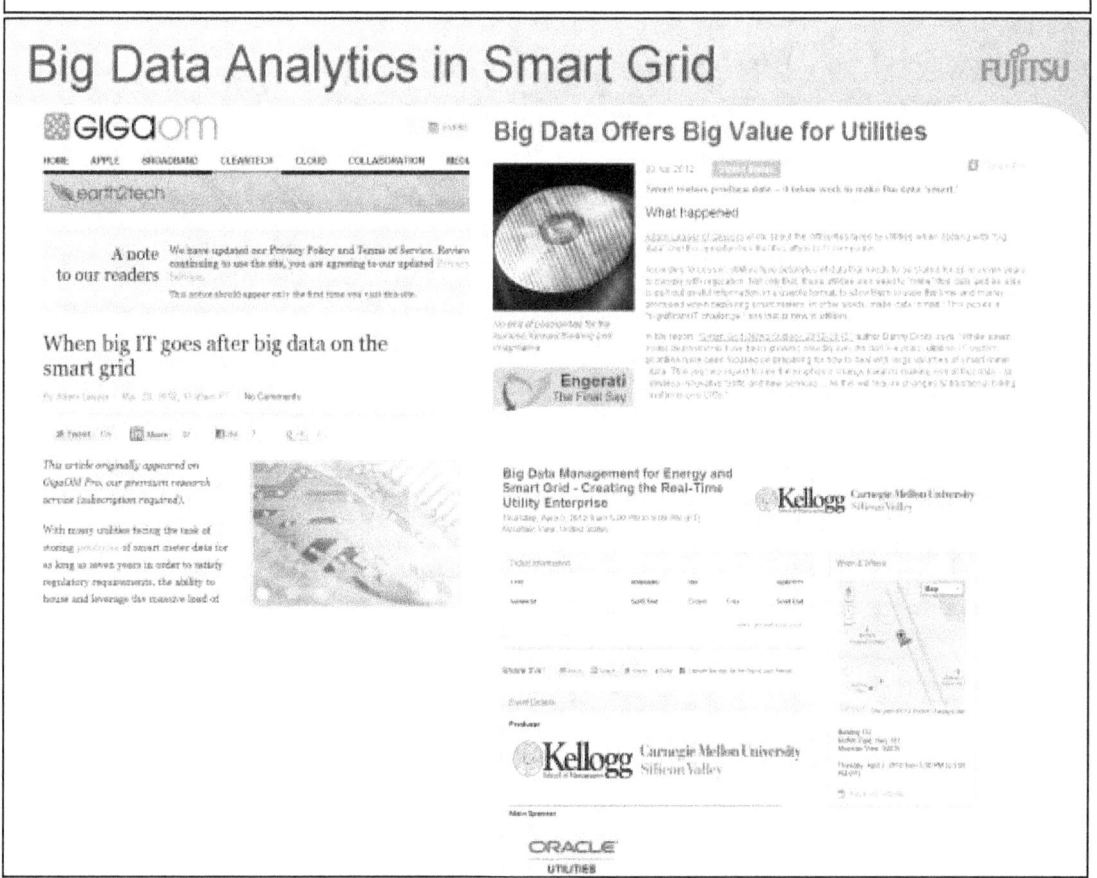

CSA Created Working Group on Big Data — FUJITSU

- Fujitsu is chairing the working group

- Please consider contributing

CSA *cloud security alliance℠* **Big Data Working Group**

Proposed Charter

April 2012

Case Study: Detection of Electricity Theft — FUJITSU

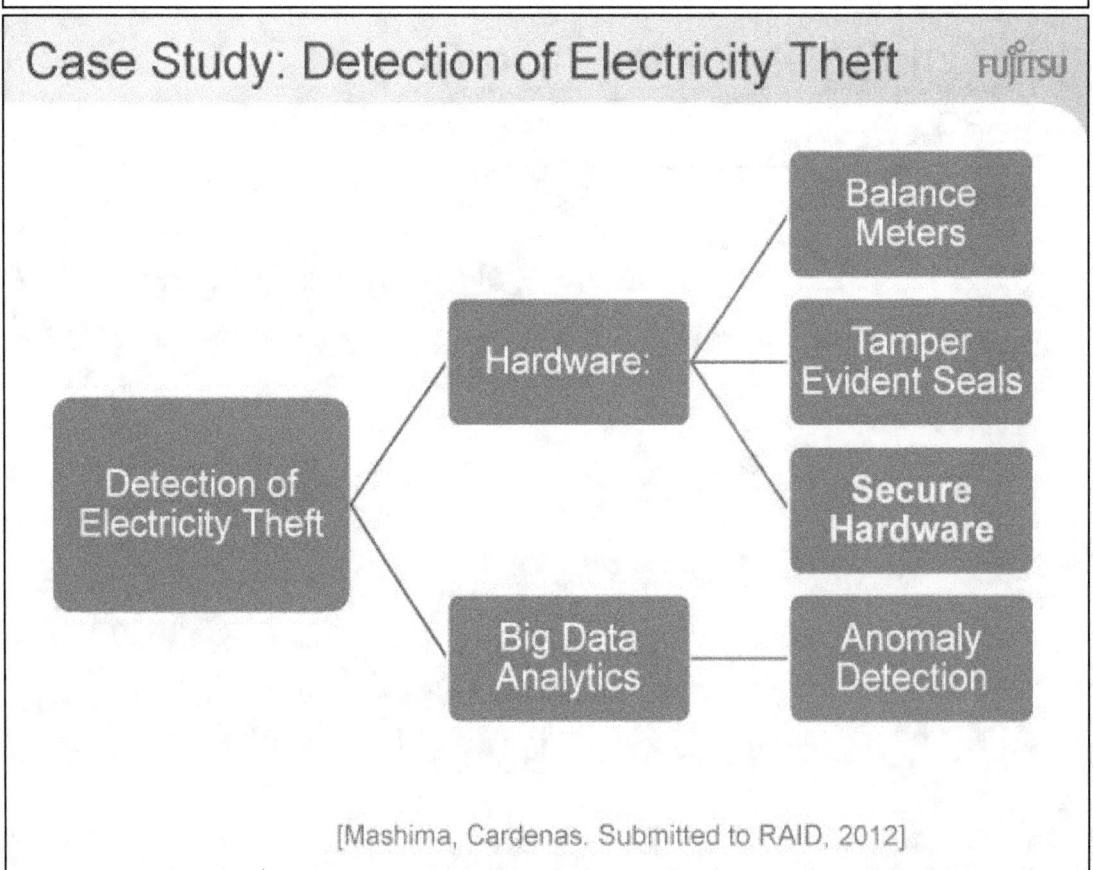

[Mashima, Cardenas. Submitted to RAID, 2012]

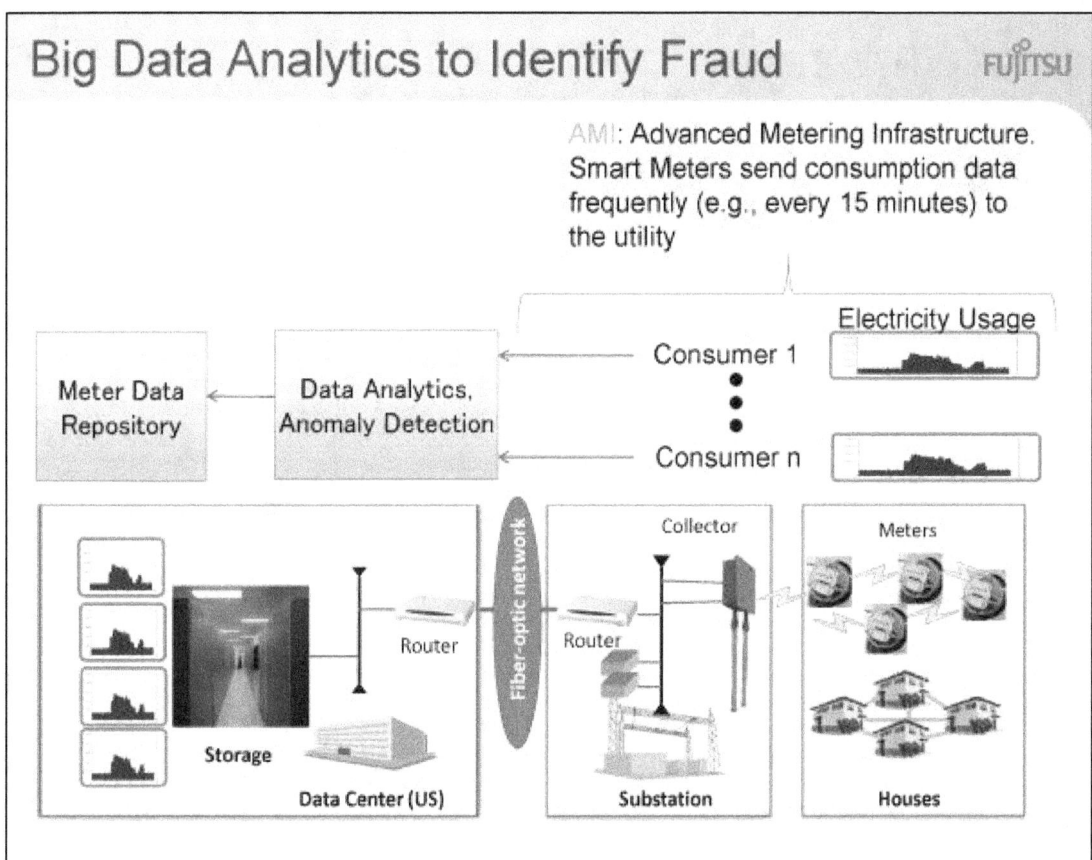

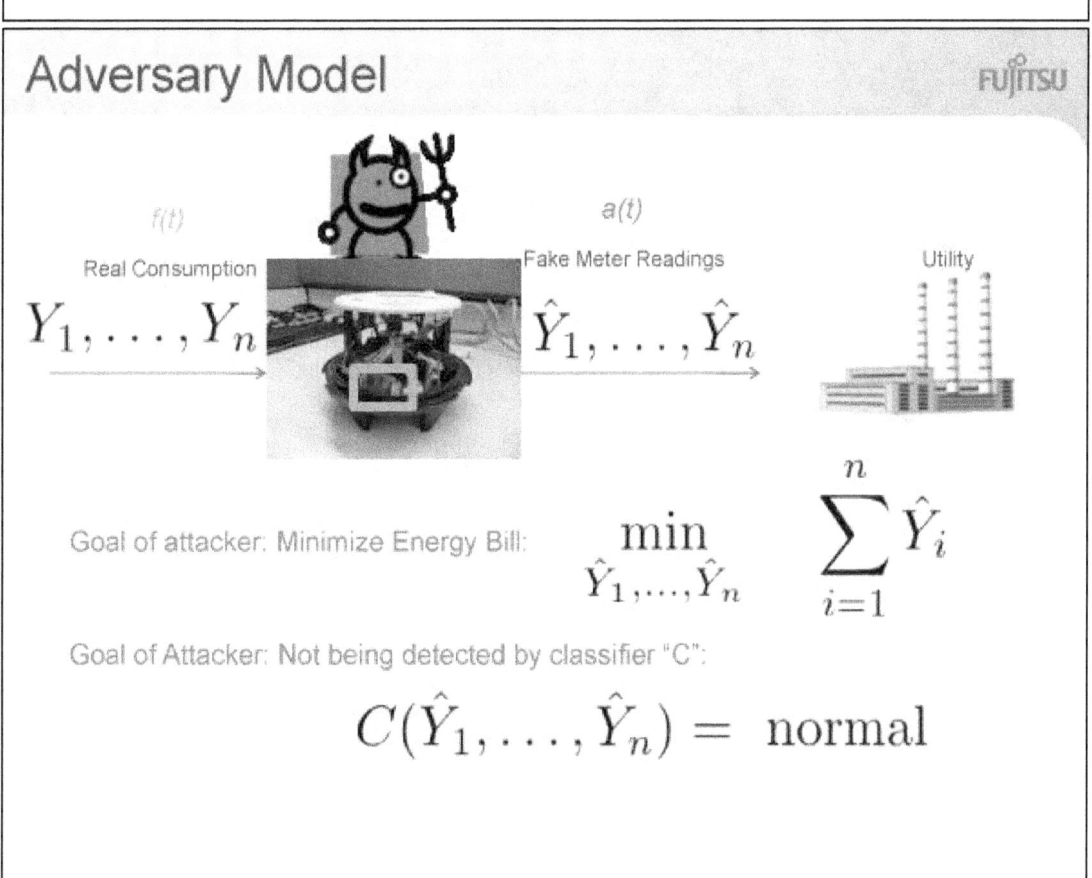

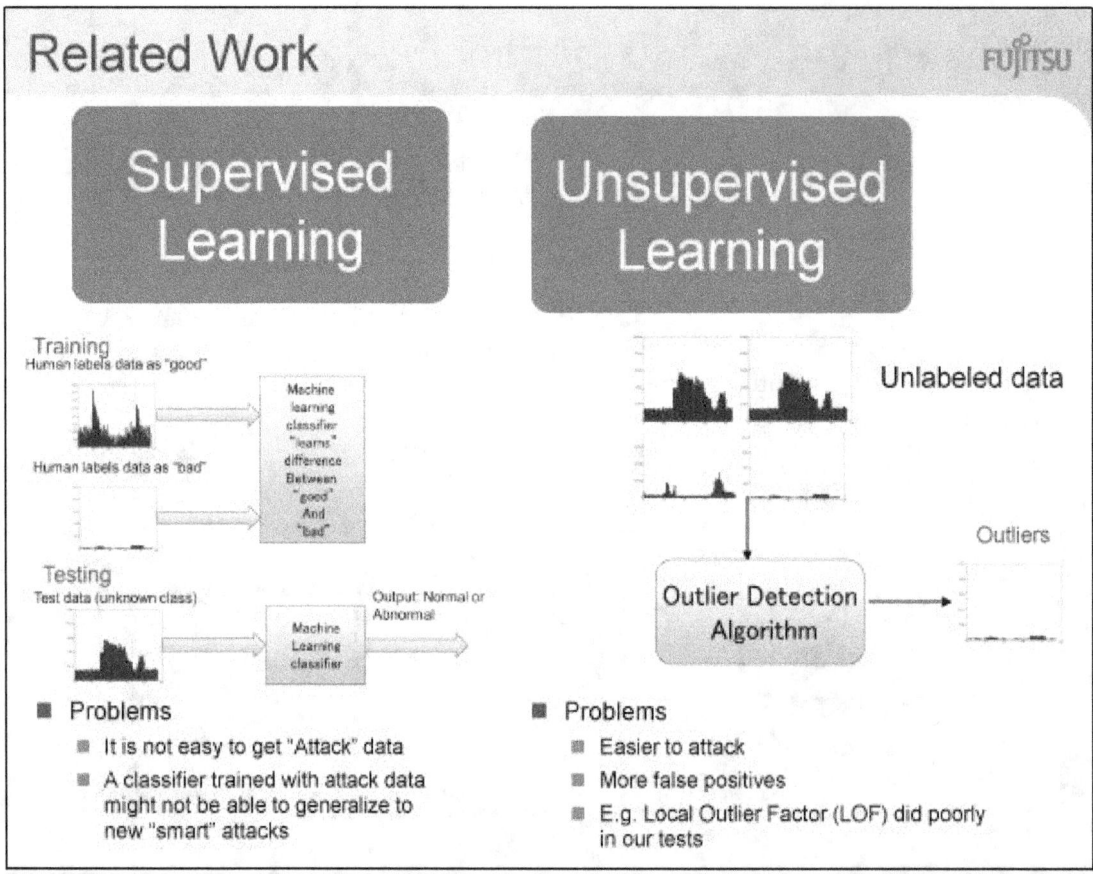

Related Work — FUJITSU

Supervised Learning

Training
Human labels data as "good"

Human labels data as "bad"

Machine learning classifier "learns" difference Between "good" And "bad"

Testing
Test data (unknown class)

Machine Learning classifier

Output: Normal or Abnormal

- Problems
  - It is not easy to get "Attack" data
  - A classifier trained with attack data might not be able to generalize to new "smart" attacks

Unsupervised Learning

Unlabeled data

Outlier Detection Algorithm

Outliers

- Problems
  - Easier to attack
  - More false positives
  - E.g. Local Outlier Factor (LOF) did poorly in our tests

New Idea: — FUJITSU

- We only have "good" data
  - Do not assume we have access to "attack" data
  - Train only one class ("good" class)
- We have prior knowledge of attack invariant
  - We know attackers want to lower energy consumption
  - Include this information for the "bad" class
- Composite Hypothesis Testing formulation:

$$H_0 : P_0$$
$$H_1 : P_\gamma \text{ s.t. } \mathbb{E}_\gamma[Y] < \mathbb{E}_0[Y]$$

$$Y_{k+1} = \sum_{i=1}^{p} A_i Y_{k-i} + \sum_{j=0}^{q} B_j (V_{k-j} + \theta)$$
under $H_0 : \theta = 0$ and under $H_1 : \theta = -\gamma, \gamma > 0.$

Problem: We Do Not Have Positive Examples

- Because meters were just deployed, we do not have examples of "attacks"

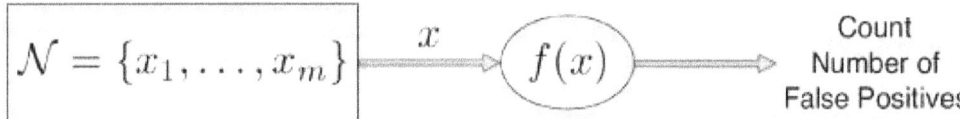

Negative Examples   Binary Classifier

$$\mathcal{N} = \{x_1, \ldots, x_m\}$$

x $f(x)$ Count Number of False Positives

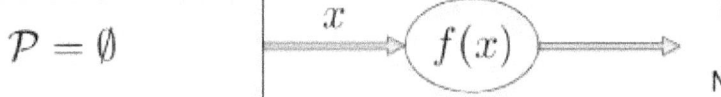

Positive Examples   Binary Classifier

$$\mathcal{P} = \emptyset$$

x $f(x)$ No Metric for False Negatives?

Our Proposal: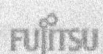

- Find the worst possible undetected attack for each classifier, and then find the cost (kWh Lost) of these attacks

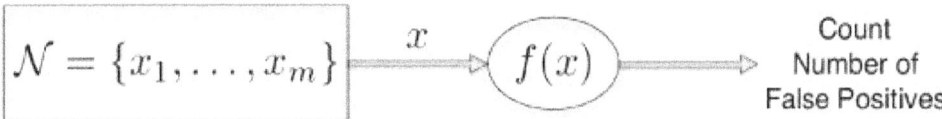

Negative Examples   Binary Classifier

$$\mathcal{N} = \{x_1, \ldots, x_m\}$$

x $f(x)$ Count Number of False Positives

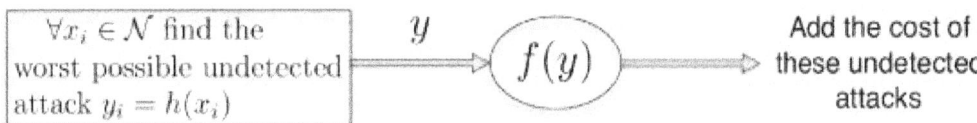

Positive Examples   Binary Classifier

$\forall x_i \in \mathcal{N}$ find the worst possible undetected attack $y_i = h(x_i)$

y $f(y)$ Add the cost of these undetected attacks

Evaluation

FUJITSU

- We tried many anomaly detectors
  - Average
  - CUSUM
  - EWMA
  - LOF
  - ARMA-GLR
- ARMA GLR is the best detector:
  - For the same false positive rage, it minimizes the ability of an attacker to create undetected attacks

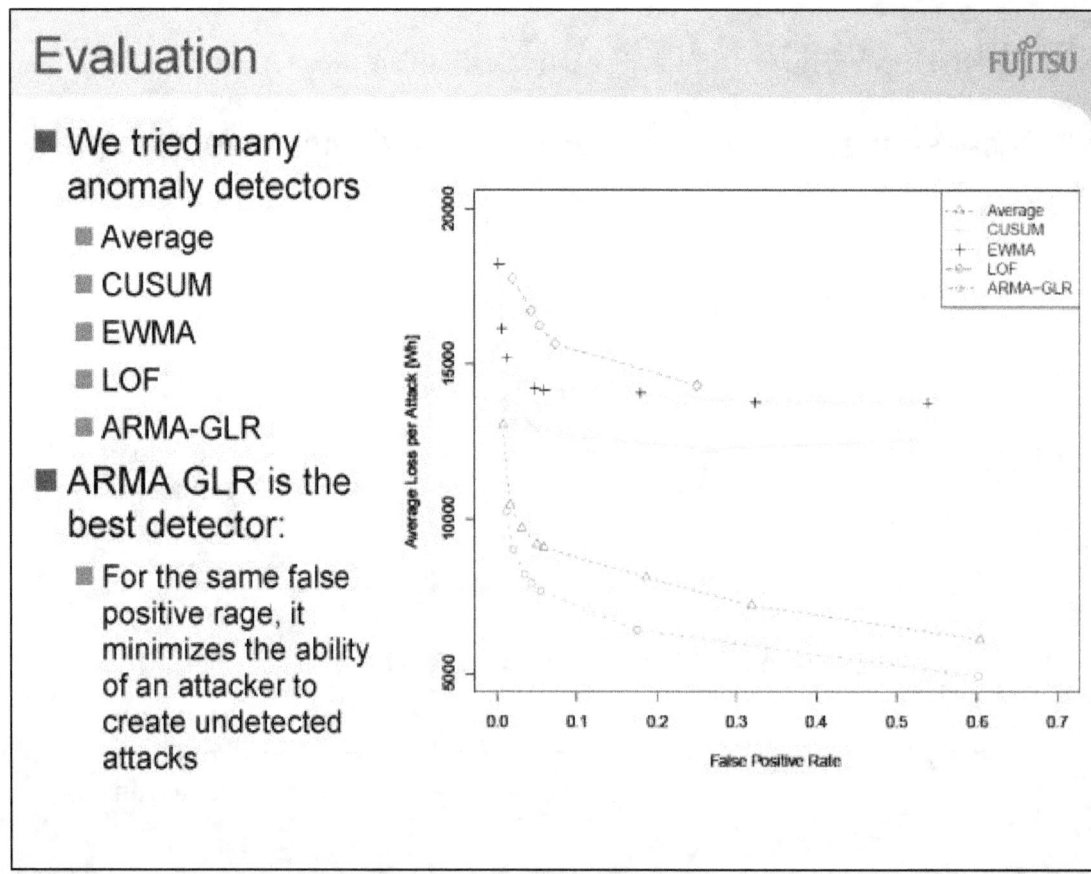

Preventing Poisoning Attacks

FUJITSU

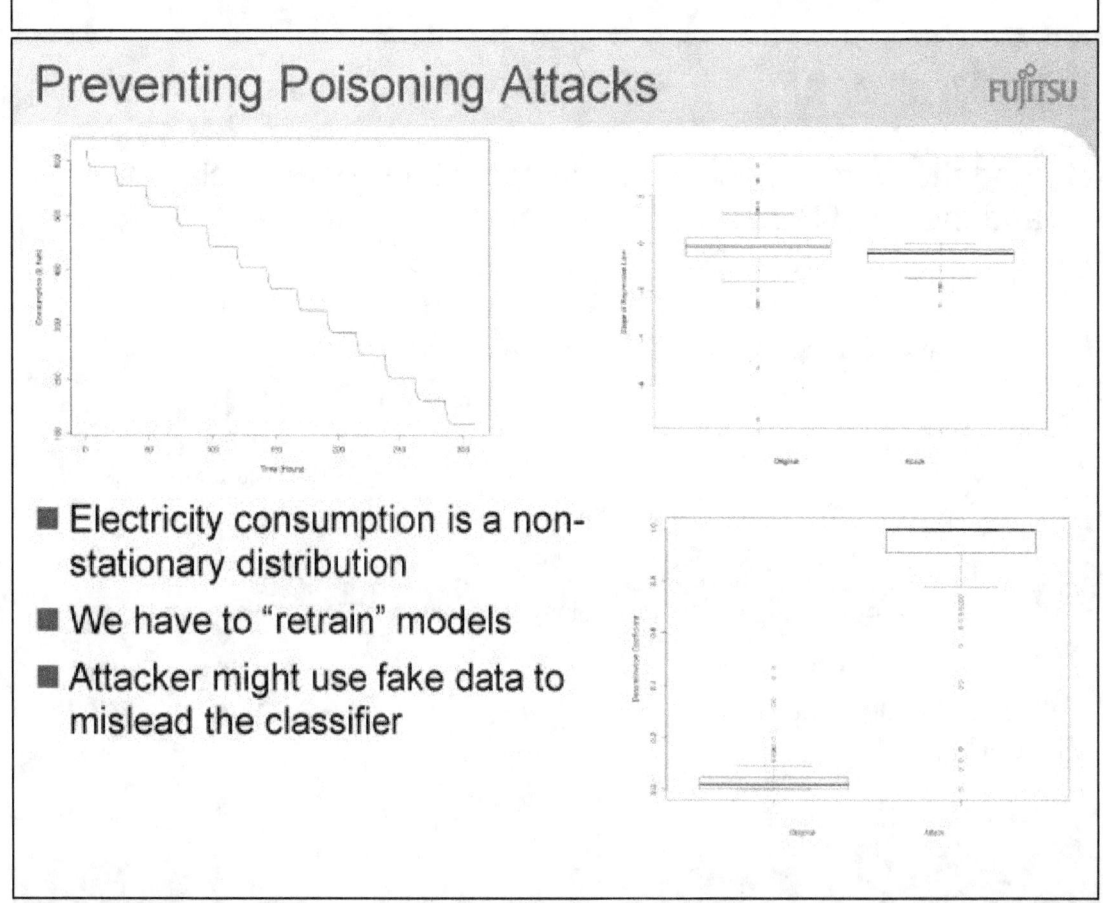

- Electricity consumption is a non-stationary distribution
- We have to "retrain" models
- Attacker might use fake data to mislead the classifier

Ongoing Work

FUJITSU

- Use in production system, experience and feedback
- Detecting other anomalies.

Normal Consumption Profile

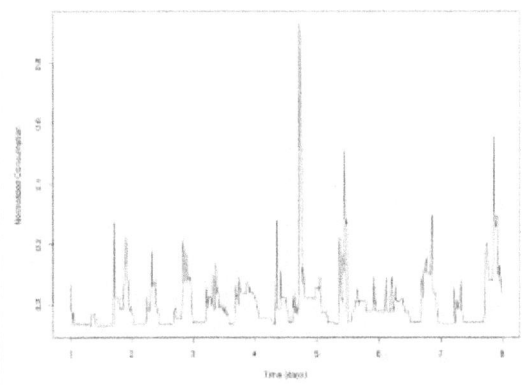

Abnormal Consumption Profile

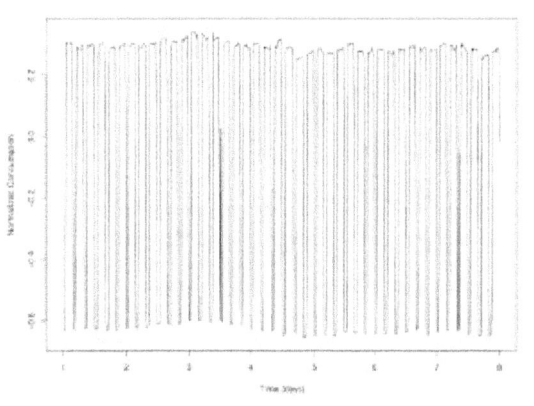

Three Steps to Improve CPS Security

FUJITSU

- Short Term
  - Incentives
  - Software reliability
  - Solve basic vulnerabilities
- Medium Term
  - Leverage Big Data for Situational Awareness
- Long Term Research
  - Resilient estimation and control algorithms

205

Previous Work in Security: What can Help in Securing CPS?

FUJITSU

- Prevention
  - Authentication, Access Control, Message Integrity, Software Security, Sensor Networks
- Detection
- Resiliency
  - Separation of duty, least privilege principle
- Incentives for vendors and asset owners to implement security best practices

Previous Work in Security: What is Missing for Secure CPS?

FUJITSU

- What is new and fundamentally different in control systems security?
- Model interaction with the physical world
  - How can the attacker manipulate the physical world?

- Attacks to Regulatory Control
  - A1 and A3 are deception attacks: the integrity of the signal is compromised
  - A2 and A4 are DoS attacks
  - A5 is a physical attack to the plant

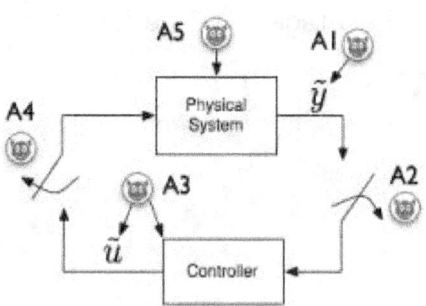

Safety Mechanisms do not Work Against Attacks

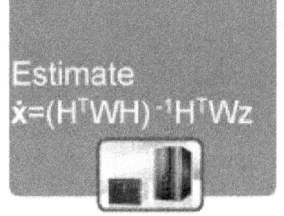

- Fault-Detection Algorithms do not Work Against Attackers
  - Liu, Ning, Reiter. CCS 09
- Attacks are different than failures!
  - Non-correlated, non-independent, etc.
- Their study is missing:
  - Impact (risk assessment) of attacks?
  - Countermeasures?

CPS security is different from IT and Control Systems Safety/Fault Detection

- So security is important; but are there new research problems, or can the problems be solved with
  - Traditional IT security? AC, IDS, AV, Separation of duty, least priv. etc.
  - Control Algorithms? Robust control, fault-tolerant control, safety, etc.
- Missing in IT Security
  - Understanding effects in the physical world
  - Attacker strategies
  - Attack detection algorithms based on sensor measurements
  - Attack-resilient estimation and control algorithms
- Missing in Control
  - Realistic attack models
  - Failures are different from Attacks!
    - Liu et.al. CCS 09, Maroochy, Stuxnet, etc.
  - Argument: Robust Control + IT Security => Resilient CPS

New CPS Research Directions

FUJITSU

- Threat assessment:
  - How to model attacker and his strategy
  - Consequences to the physical system
- Attack-resilient control algorithms
  - CPS systems that degrade gracefully under attacks
- Attack-detection by using models of the physical system
  - Study stealthy attacks (undetected attacks)
- Big Data Analytics
  - Situational awareness
- Privacy
  - Privacy-aware CPS algorithms

Papers articulating new research for CPS security
Cardenas, Amin, Sastry, HotSec 08, & ICDCS Workshop (08)

GAO Agrees: We Need new Research for CPS Security

FUJITSU

"Recommendations"

NIST and FERC should coordinate the development and adoption of smart grid guidelines and standards

EISA 2007

NIST
- SGIP CSWG
- NIST-IR 7628

FERC
- NERC CIP

Bulk Power System Regulation!

GAO Review 2011

NIST missing CPS Security

Requirements for Secure Control

FUJITSU

- Step 1: Threat Model/Assessment
  - Identify requirements
- Traditional Security Requirements: CIA (Confidentiality, Integrity, Availability)
  - What are the requirements of secure control?

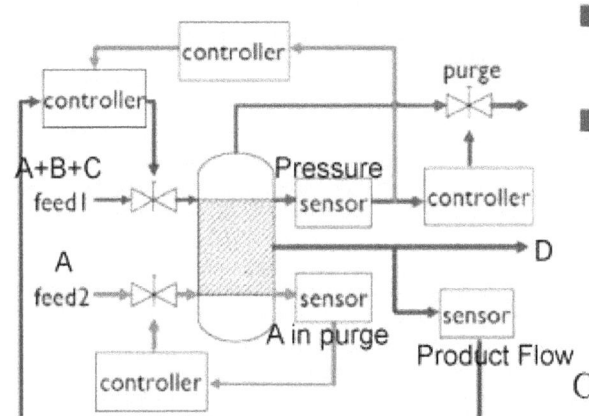

- **Safety Constraint:**
  - Pressure < 3000kPa
- **Operational Goal:**
  - Cost:
    - Proportional to the quantity of A and C in purge,
    - Inversely proportional to the quantity of the final product D

$$\text{Cost} = \frac{F_3}{F_4}(2.206 y_{A3} + 6.177 y_{C3})$$

[Journal of Critical Infrastructure Protection 2009]

Not all Compromises affect Safety

FUJITSU

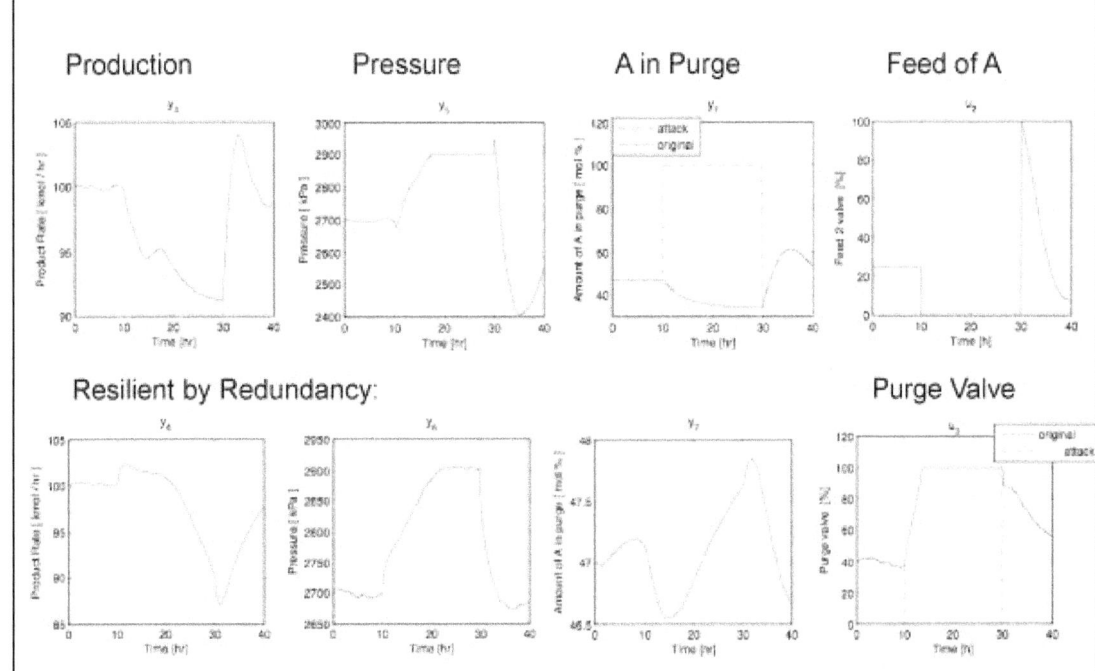

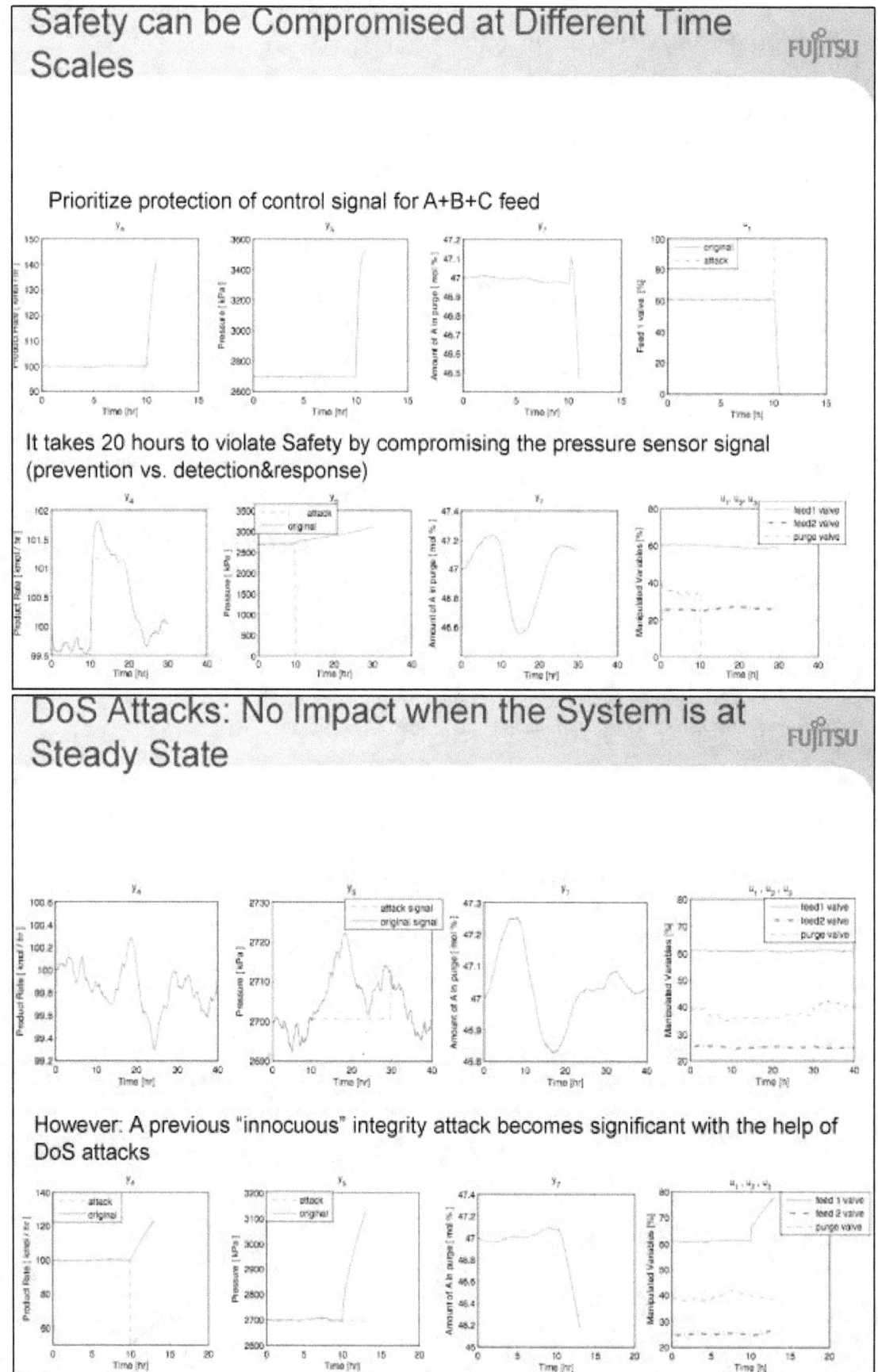

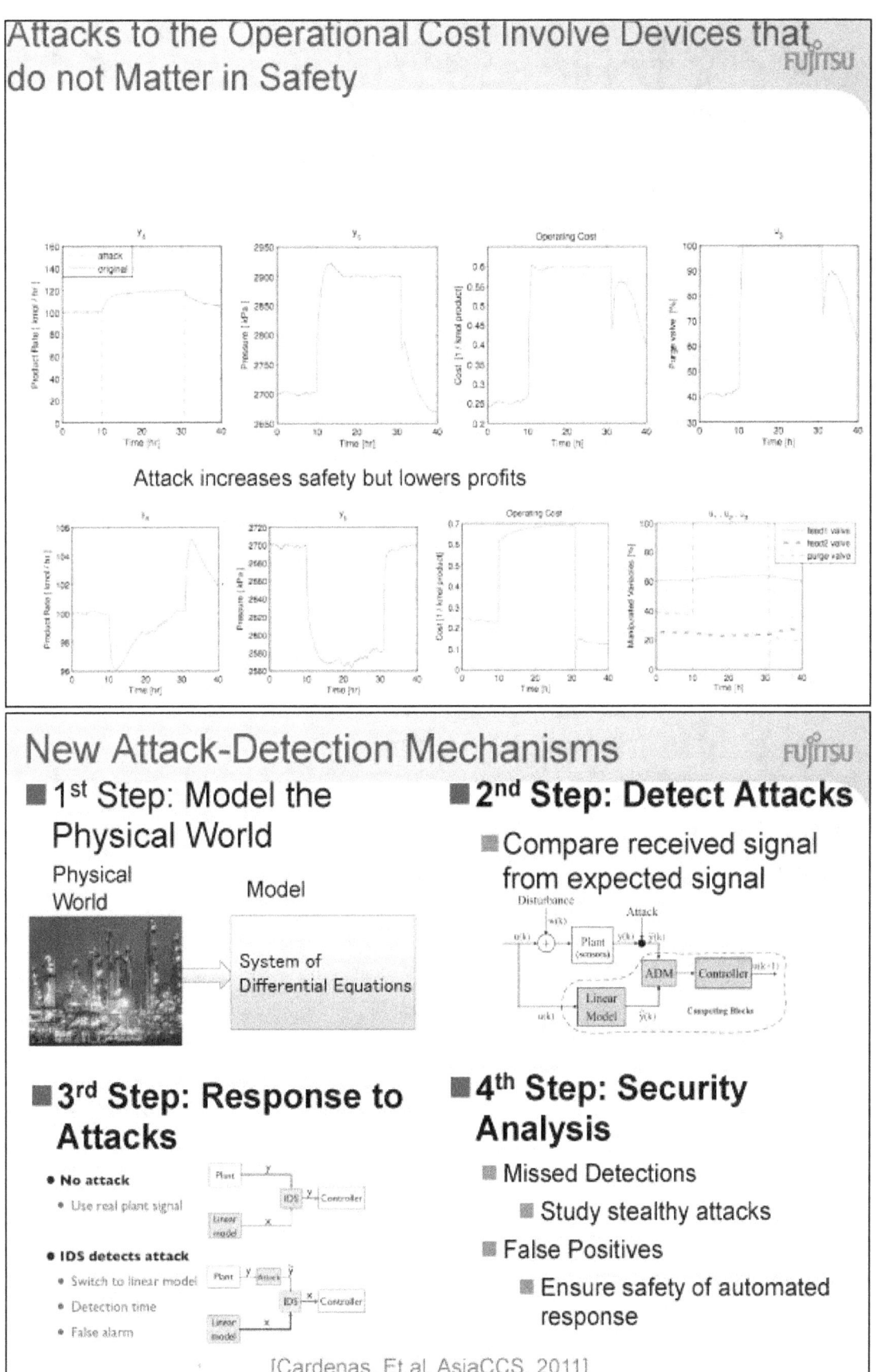

Attacker Strategy: Stealthy Attacks

FUJITSU

- ■ Attacker
  - ■ Knows our detection model and its parameters
  - ■ Wants to be undetected for n time steps
  - ■ Wants to maximize the pressure in the tank
- ■ Surge attack
$$\tilde{y}_K = \begin{cases} y^{min} & \text{if } S_{k+1} \leq \tau \\ \hat{y}_K - |\tau + b - S_k| & \text{if } S_{k+1} > \tau \end{cases}$$

- ■ Bias attack
$$\tilde{y}_k = \hat{y}_k - (\tau/n + b)$$

- ■ Geometric attack $\tilde{y}_k = \hat{y}_k - \beta\alpha^{n-k}$

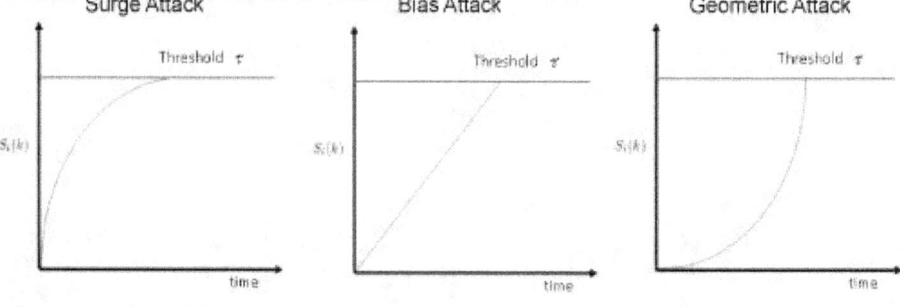

Impact of Undetected Attacks

FUJITSU

- ■ Even geometric attacks cannot drive the system to an unsafe state
- ■ If an attacker wants to remain undetected, she cannot damage the system

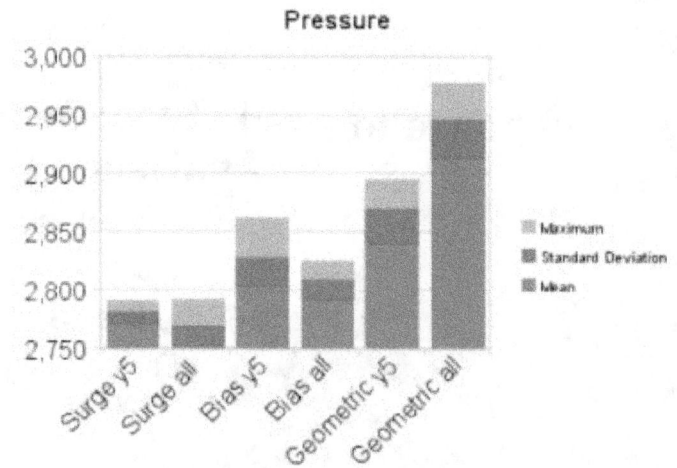

Control Resilient to DoS Attacks

FUJITSU

For constrained linear systems

$$x_{k+1} = Ax_k + Bu_k^a + w_k, \qquad k = 1, \ldots, N-1$$

$$x_k^a = \gamma_k x_k, \; u_k^a = \nu_k u_k, \qquad (\gamma_k, \nu_k) \in \{0, 1\}^2$$

find causal feedback policies $u_k = \mu_k(x_0^a, \ldots, x_k^a)$, that
minimize $J(x_0, \mathbf{u}, \mathbf{w}) = \sum_{k=1}^{N} x_k^\top Q^{xx} x_k + \sum_{k=1}^{N-1} \nu_k u_k^\top Q^{uu} u_k$,
subject to power constraints

$$\begin{pmatrix} x_k^a \\ u_k^a \end{pmatrix}^\top \begin{pmatrix} H_i^{xx} & 0 \\ 0 & H_i^{uu} \end{pmatrix} \begin{pmatrix} x_k^a \\ u_k^a \end{pmatrix} \leq \beta_i, \qquad i = 1, \ldots, L_1.$$

and safety constraints

$$\begin{pmatrix} x_k^a \\ u_k^a \end{pmatrix} \in \mathcal{T}_j, \qquad j = 1, \ldots, L_2.$$

for all disturbances $\mathbf{w} \in \mathbf{W}_\alpha$ OR $\mathbf{w} \sim \mathcal{N}(0, W)$ and a given set of
$(\gamma_0^{N-1}, \nu_0^{N-1}) \in \mathcal{A}_{pq}$ attack signatures.

[Amin, Cardenas, Sastry. HSCC / CPSWeek 2009]

Privacy-Preserving Control

FUJITSU

- **Data Minimization Principle**
  - How much data do we really need to collect for accurate estimation/control?
  - Quantity: sampling
  - Quality: quantization
- **Demand Response (DR)**

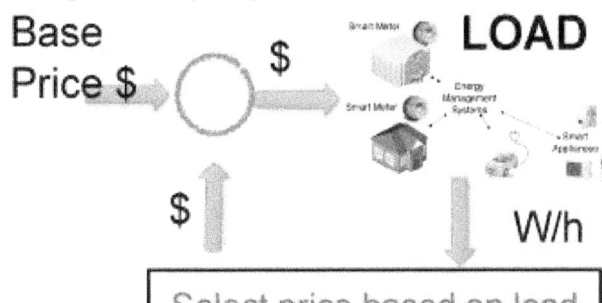

Base Price $

$

LOAD

$

W/h

Select price based on load
(and available supply)

[Cardenas, Amin, Schwartz. HiCoNS / CPSWeek 2012]

CPS Research for Smart Grid

FUJITSU

- DoE 2020 Vision:
  - Maintain Smart Grid functions under attack
- Develop resilient algorithms for:

Smart Grid Function	• Untrusted input
State Estimation	• Power flow sensors
Network Topology Processor	• Breakers
Electricity Markets	• Prices
Load balancing	• Smart meter, control data
Transmission/Distribution Automation	• Flexible Alternate Current Transmission System (FACTS)

13. State Estimation and Contingency Analysis of the Power Grid in a Cyber-Adversarial Environment

Robin Berthier[1], Rakesh Bobba[1], Matt Davis[2], Kate Rogers[1, 2], and Saman Zonouz[3]

| [1]Information Trust Institute University of Illinois at Urbana-Champaign Urbana, IL, USA {rgb, rbobba}@illinois.edu | [2]PowerWorld Corporation Champaign, IL, USA matt@powerworld.com, krogers6@illinois.edu | [3]Department of Electrical and Computer Engineering University of Miami Miami, USA s.zonouz@miami.edu |

Abstract—Contingency analysis is a critical activity in the context of the power infrastructure, because it provides a guide for resiliency and enables the grid to continue operating even in the case of failure. A critical issue with the current evolution of the power grid into a so-called smart grid is the introduction of cyber-security threats due to the pervasive deployment of communication networks and digital devices. In this paper, we introduce a cyber-physical security evaluation technique to take into account those threats. The goal of this approach is to augment traditional contingency analysis by not only planning for accidental contingencies but also for malicious compromises. This solution requires a new unified formalism to model the whole cyber-physical system including interconnections among the cyber and physical components. The system model is later used to assess potential impacts of both cyber and physical contingencies in order to prioritize prevention and mitigation efforts.

Keywords-component; formatting; style; styling; insert (key words)

I. INTRODUCTION

State estimation and contingency analysis are the two most fundamental tools for monitoring the power system. State estimation is the process of fitting data coming in from sensors in the field to a system model and determining an estimate of the power system state. By its nature, state estimation depends on the communication infrastructure, commonly called the SCADA (system control and data acquisition) system. These systems are currently undergoing many changes as new sensors and communications infrastructure is being deployed as part of the smart grid initiative. Indeed, the smart grid becomes the perfect example of a large and complex cyber-physical system.

This complexity and the inter-connected nature of the power grid infrastructure introduce critical cyber security threats that can impact state estimation and contingency analysis at multiple levels. First, cyber attacks can breach the integrity of sensor data required for state estimation. Second, adversaries can initiate incidents that are out of the scope of traditional reliability analysis, such as cascading failures that could not be caused by accident.

While the problem of detecting and mitigating cyber intrusions has been extensively studied over the past two decades in the context of traditional IT systems, the requirements and constraints of a cyber-physical system such as the smart grid are different and usually more stringent. For example, a lot of power grid components have timing requirements that prevent traditional security solution from being deployed. Moreover, the fact that cyber systems and power grid components are inter-connected creates a new set of dependencies for which the security community has currently a poor understanding. Recently, several attempts have been conducted to model and analyze the cyber-physical threats in an offline manner [1—4]; however, to the best of our knowledge, there has been no efficient online solution proposed for cyber-physical attack detection and contingency analysis.

In this work, we present a cyber-physical contingency analysis framework that takes into account cyber- and power-side network topologies, malicious cyber asset compromises and power component failures. In particular, during an offline process, the cyber network topology and global access control policies is analyzed automatically to generate a network connectivity map that represents a directed graph encoding inter-host accessibilities. The resulting connectivity map is then used to generate a Markovian state-based model of the power-grid in an online manner. At any time instance, the current security state can be estimated using the generated model and the triggered set of cyber-side intrusion detection sensor alerts. Using a new cyber-physical security index, the criticality level of any system state is measured and a ranked list of potential cyber and/or physical contingencies that needs to be taken care of in priority is produced.

II. EXAMPLE

To illustrate our approach, we present preliminary results on the case study of a power grid infrastructure that is based on a real-world power control network. Figure 1 shows the cyber-side topology, i.e., power control network topology of the power grid. Figure 2 shows the physical power system topology. As illustrated in Figure 1, the computer systems (gray circles identified by IP addresses) are interconnected through routers (blue circles) and network firewalls (red circles). The network topology is an abstract version of a real-world power control network. It is initially assumed that the attackers reside on a remote computer system denoted by the node labeled *Internet* in Figure 1. Figure 2 shows how the power system generation and load buses are interconnected through transmission lines. We use the NetAPT network analysis tool to parse and analyze the access control policies of the network and generate automatically the power grid attack graph that enumerates all possible attack paths against cyber assets and physical power system components. The attack graph for this case study is shown in Figure 3. Each node in the attack graph represents a compromised or damaged cyber or physical asset within the power grid.

The graph from Figure 3 provides the structure on which we can run state estimation algorithms to assess, at each time instant, the current state of the power grid given the past sequence of measurements from power system sensors (e.g., phase measurement units) and the cyber side security sensors (e.g., intrusion detection systems). In addition, this graph can be used to empirically evaluate the impact of cyber-physical contingency by taking into account what the attackers could or would do from any state of the power grid. This structure provides the power system operators with an invaluable knowledge base regarding global impacts of various cyber network or power system contingencies that can assist the identification of the parts of the power grid that need to be the focus of protection and monitoring efforts.

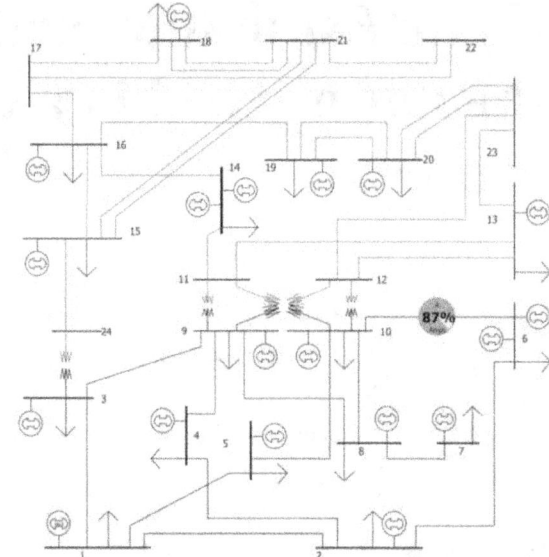

Figure 2: Power system topology

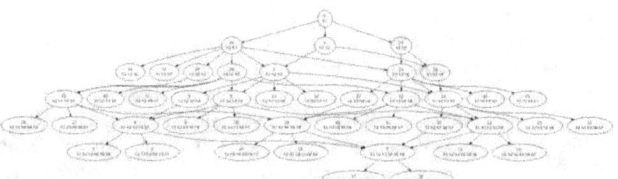

Figure 3: Attack graph

III. CONCLUSION

In summary, the main contribution of this work is to introduce a new framework for cyber-physical contingency analysis that addresses the challenge of state estimation of a complex and large-scale cyber-physical system. Next steps on this research 1) include the implementation and the evaluation of efficient state estimation algorithms that can cope with the large state space in a timely manner; 2) the introduction of a probabilistic solution to identify and ignore noisy or maliciously corrupted measurements among the sensory data; and 3) the capability to make predictions under high level of uncertainty.

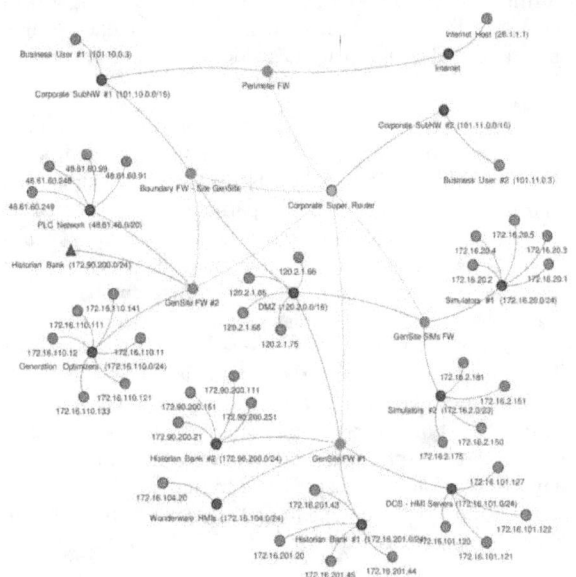

Figure 1: Control network configuration

REFERENCES

[1] Thomas M Chen, Senior Member, Juan Carlos Sanchez-aarnoutse, and John Buford. Petri net modeling of cyber-physical attacks on smart grid. *Smart Grid IEEE Transactions on*, 2(99):1–9, 2011.

[2] Y. Mo, T. H.-J. Kim, K. Brancik, D. Dickinson, H. Lee, A. Perrig, and B. Sinopoli. Cyber-physical security of a smart grid infrastructure. *Proceedings of the IEEE*, 100(1):195–209, jan. 2012.

[3] Fabio Pasqualetti, Florian Do'rfler, and Francesco Bullo. Cyber-physical attacks in power networks: Models, fundamental limitations and monitor design. *CoRR*, abs/1103.2795, 2011.

[4] Siddharth Sridhar, Adam Hahn, and Manimaran Govindarasu. Cyber-physical system security for the electric power grid. *Proceedings of the IEEE*, 100(1):210–224, 2012.

State Estimation and Contingency Analysis of the Power Grid in a Cyber-Adversarial Environment

Robin Berthier[1], Rakesh Bobba[1], Matt Davis[2], Kate Rogers[2], and Saman Zonouz[3]

[1]Information Trust Institute
University of Illinois at
Urbana-Champaign
Urbana, IL, USA
{rgb, rbobba}@illinois.edu

[2]PowerWorld Corporation
Champaign, IL, USA
{matt, kate}@powerworld.com

[3]Department of Electrical
and Computer Engineering
University of Miami
Miami, USA
s.zonouz@miami.edu

Motivation

- New technologies and new resources
- Extensive data integration
  - Sensory data
  - Control data
- Complex dependencies
- Stringent requirements

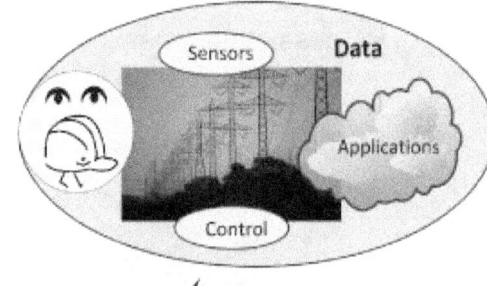

Security vs. Dependability

- Dependability and fault tolerance
  - *Accidental* failures
  - *Second party is the (unintentional) nature*
    - *Future action set can (probabilistically) be predicted*
  - *Traditional probabilistic analysis/modeling*

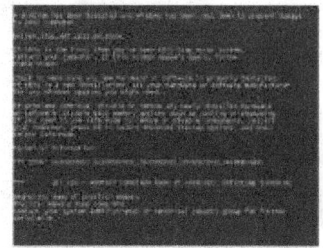

- Security and intrusion tolerance
  - *Malicious* failures
  - *Second party are (intentional) attackers*
    - *If predicted, they can exploit the prior information to damage further*
  - New solutions are needed…

3

Cyber-Physical System Security

- Systems in which cyber & physical systems are tightly integrated
  - Power systems
  - Process control networks
  - …

- (Potentially) more catastrophic security incidents…

Targeting nuclear plants

Power Control Network

4

Outline

- Power Grid Operation
  - Cyber-physical relationships
  - State estimation
- Cyber-Physical Threat Model
  - Step-1: Cyber network exploits
  - Step-2: Physical system-aware attacks
- Defense Solutions
  - Cyber network intrusion detection
  - System-aware detection and protection
    - Measurement protection and bad-data detection
  - System contingency analysis

Power Grid Operation

Cyber-physical relationships

Power System Structure

- Major components:
  - Generators: produce electricity
  - Loads: consume electricity
  - Lines (T&D): transport energy from generators to loads

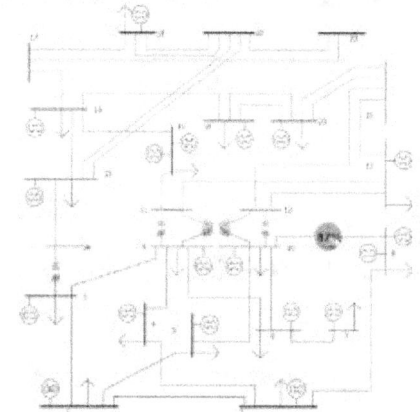

- Key Features
  - Absence of large-scale storage capabilities
  - Constraints: power balance, Kirchhoff's laws
  - Power flows through paths of "least resistance"
  - "Just-in-time" type manufacturing system

Operation and Control

- *Economics* and *reliability* are the key drivers in power system operations and control

- Economics leads to large optimization problems for
  - Resource scheduling via unit commitment
  - Least-cost dispatch of available generation

- Reliability requirements typically entail no violations of physical limits and voltages and frequencies within prescribed bounds
  - Continuous monitoring
  - Hierarchical control architecture

Monitoring and Control

- Large and complex hardware-software systems are used for real-time operations and control
  - Energy management system (EMS)
  - Supervisory control and data acquisition (SCADA)

- Frequency is closely monitored and maintained around 60 Hz
  - Area control error (ACE) is measure for frequency excursions as well as deviations from scheduled interchanges – ideally, it should be *zero*
  - Automatic generation control (AGC) implements proportional-integral-derivative (PID) control to keep ACE = *zero*

Power System Operations

Data flow in power system operations

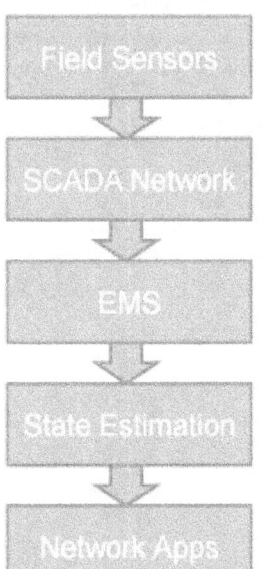

Sensors are becoming faster and more intelligent (e.g., PMUs)

SCADA networks that have traditionally been serial or microwave links are becoming network based

Network Apps include real time contingency analysis on the state estimated model

Power Grid Operation

State Estimation

Power Grid Observability

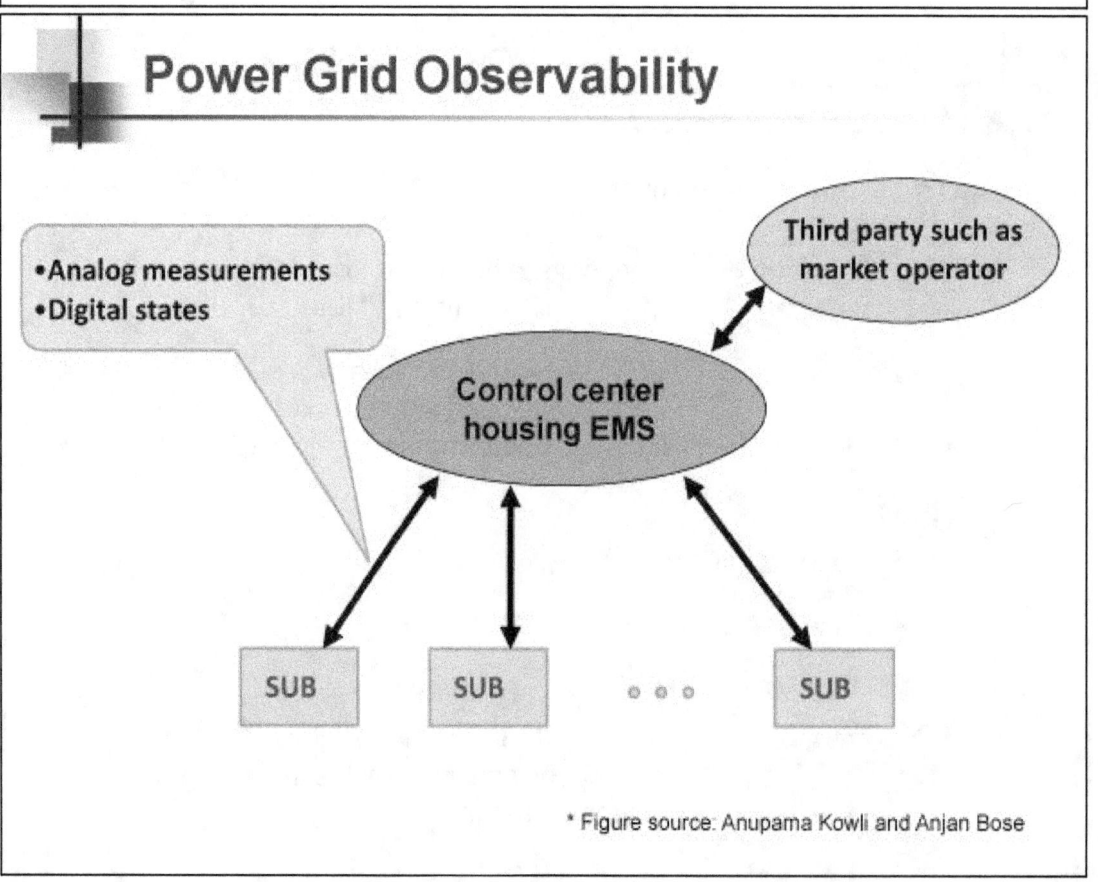

* Figure source: Anupama Kowli and Anjan Bose

State Estimation

- Key process in power system operation and control
- Problem statement: given certain measurements, find the *states* (voltages and angles) of the system

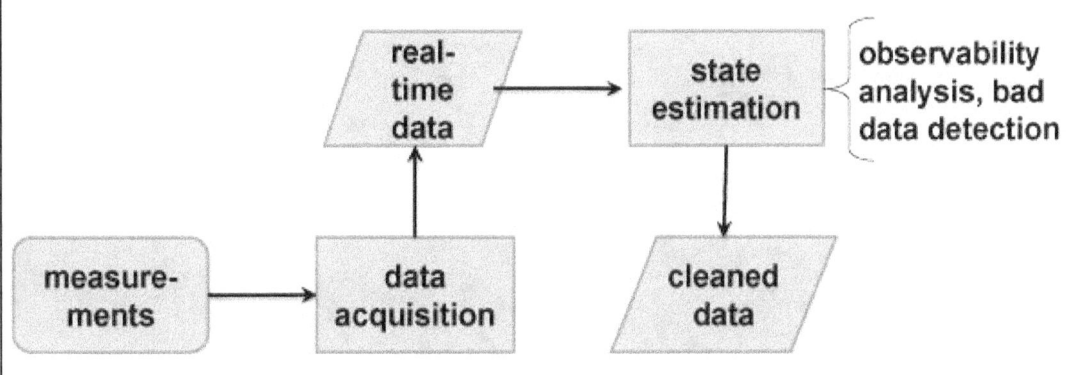

State Estimation

- The power flow is the central tool of power system planners and operators

Inputs:
System topology
Generation output
Load values

Outputs:
Voltage magnitude and angle
Line flows

$$\mathbf{P_{ij}} = \mathbf{V_i^2}[-\mathbf{G_{ij}}] + \mathbf{V_i V_j}[\mathbf{G_{ij}} \cos(\theta_i - \theta_j) + \mathbf{B_{ij}} \sin(\theta_i - \theta_j)]$$

$$\mathbf{Q_{ij}} = \mathbf{V_i^2}[-\mathbf{G_{ij}}] + \mathbf{V_i V_j}[\mathbf{G_{ij}} \sin(\theta_i - \theta_j) + \mathbf{B_{ij}} \cos(\theta_i - \theta_j)]$$

- Fundamentally, the power flow enforces the conservation of power at every Kirchoff's voltage law node in the system

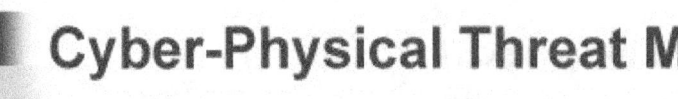

Cyber-Physical Threat Model

Step-1: Cyber network exploits

Step-2: Physical system-aware attacks

Cyber-Physical Threat

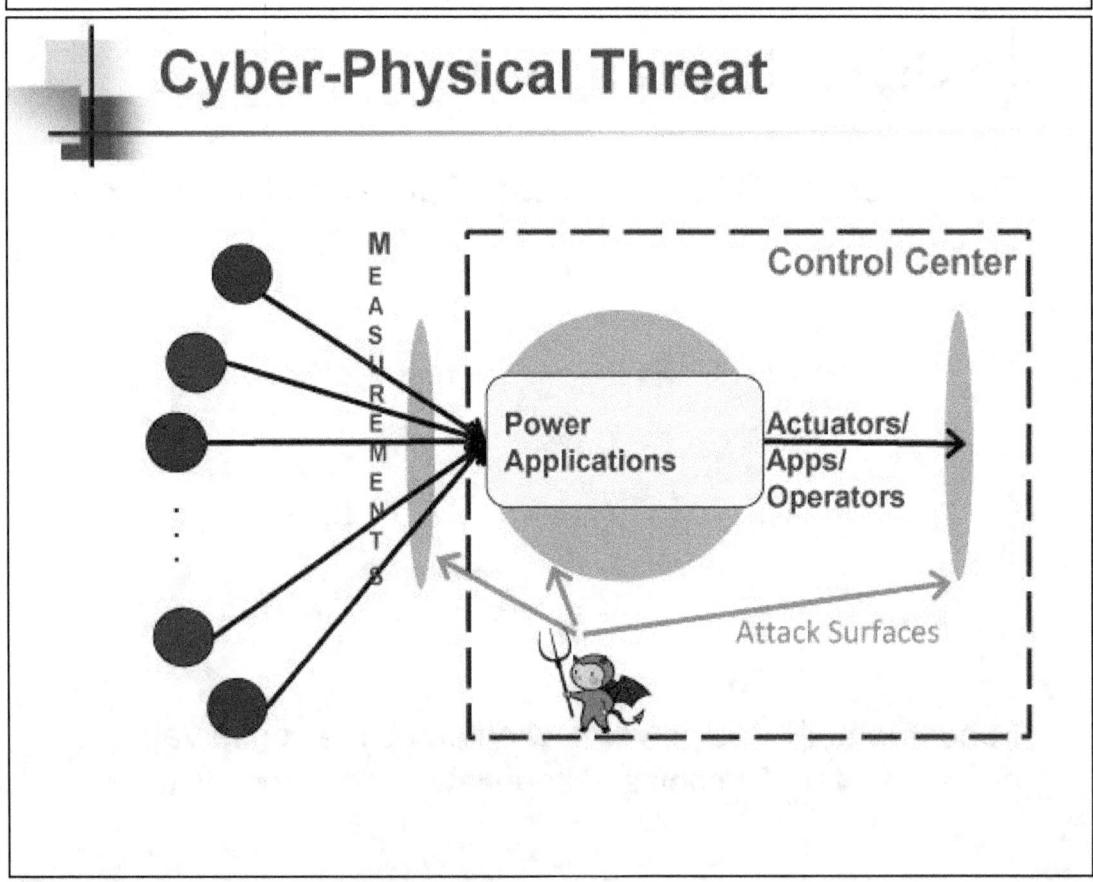

Network Exploits

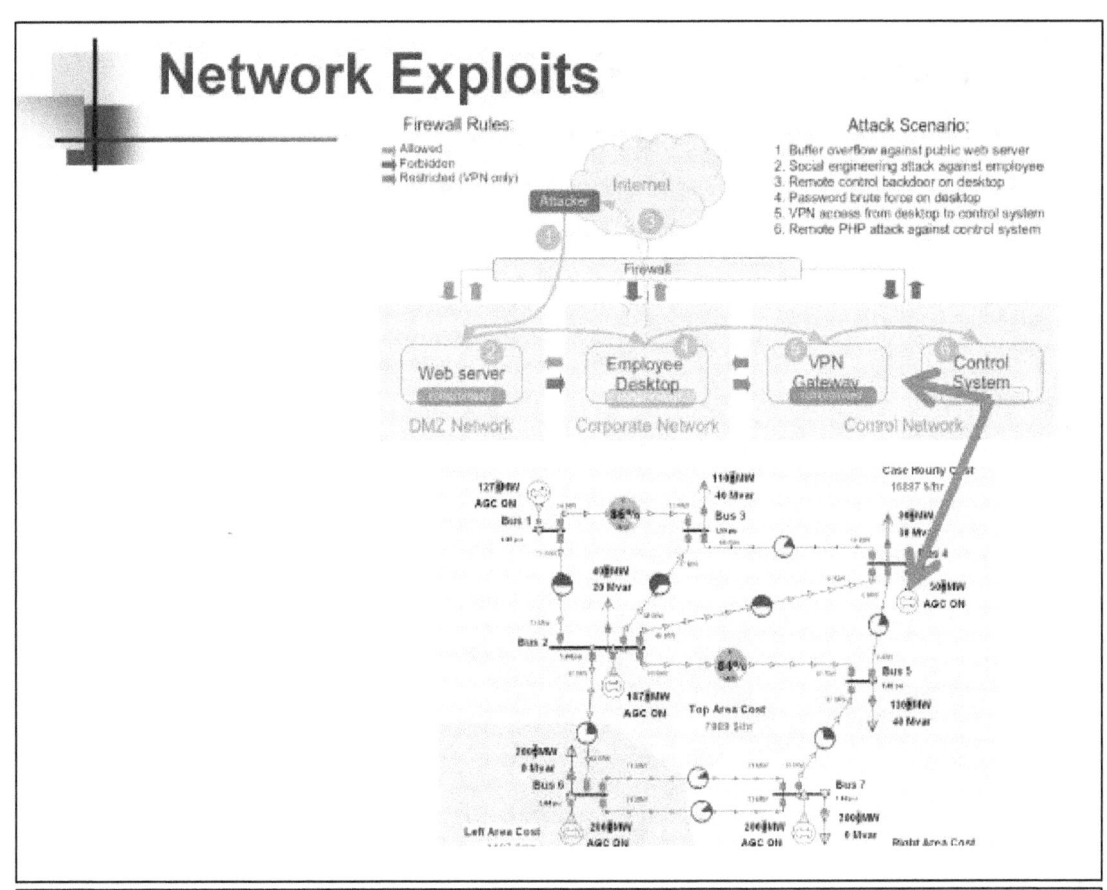

False Data Injection on State Estimation

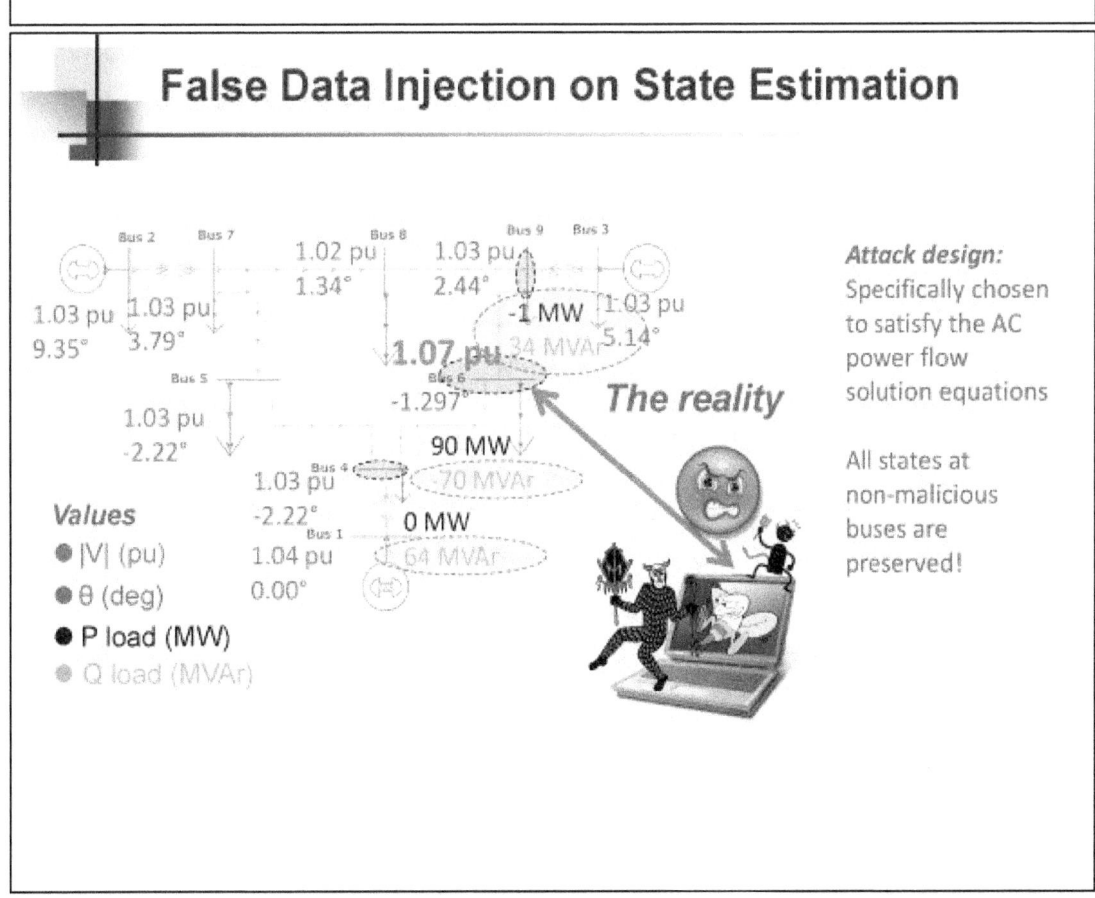

Defense Solutions

Cyber Network Intrusion Detection

Intrusion Detection Techniques

Legitimate Actions/Protocol Specification	Malicious Actions

Anomaly-based
+ detect unknown attacks
+ high scalability
- no root cause
- high false positive rate

Signature-based
+ low false positive rate
+ attack root cause
- require frequent update
- limited to known attacks

Specification-based
+ detect unknown attacks
+ high accuracy
- poor scalability
- high development cost

Specification-based Intrusion Detection

- Opportunities:
  - Leverage tight control over communication protocols and system behavior
  - Specification-based:
    - Little requirements about existing attacks
    - Ability to detect unknown attacks
    - No frequent update required
  - Enable the use of mathematical proof (formal methods)

- Challenges:
  - Scalability: stateful protocol analysis is resource intensive
  - Development costs: every protocol/application has to be specified

Solution Overview*

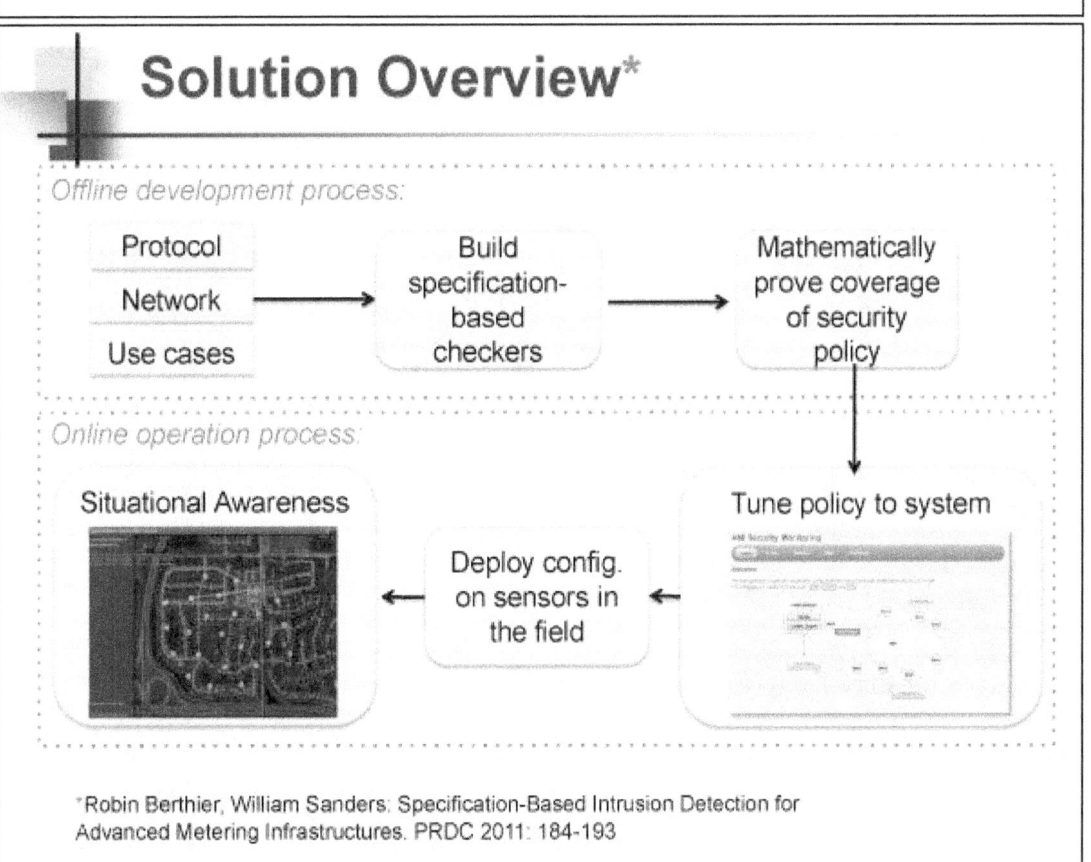

*Robin Berthier, William Sanders: Specification-Based Intrusion Detection for Advanced Metering Infrastructures. PRDC 2011: 184-193

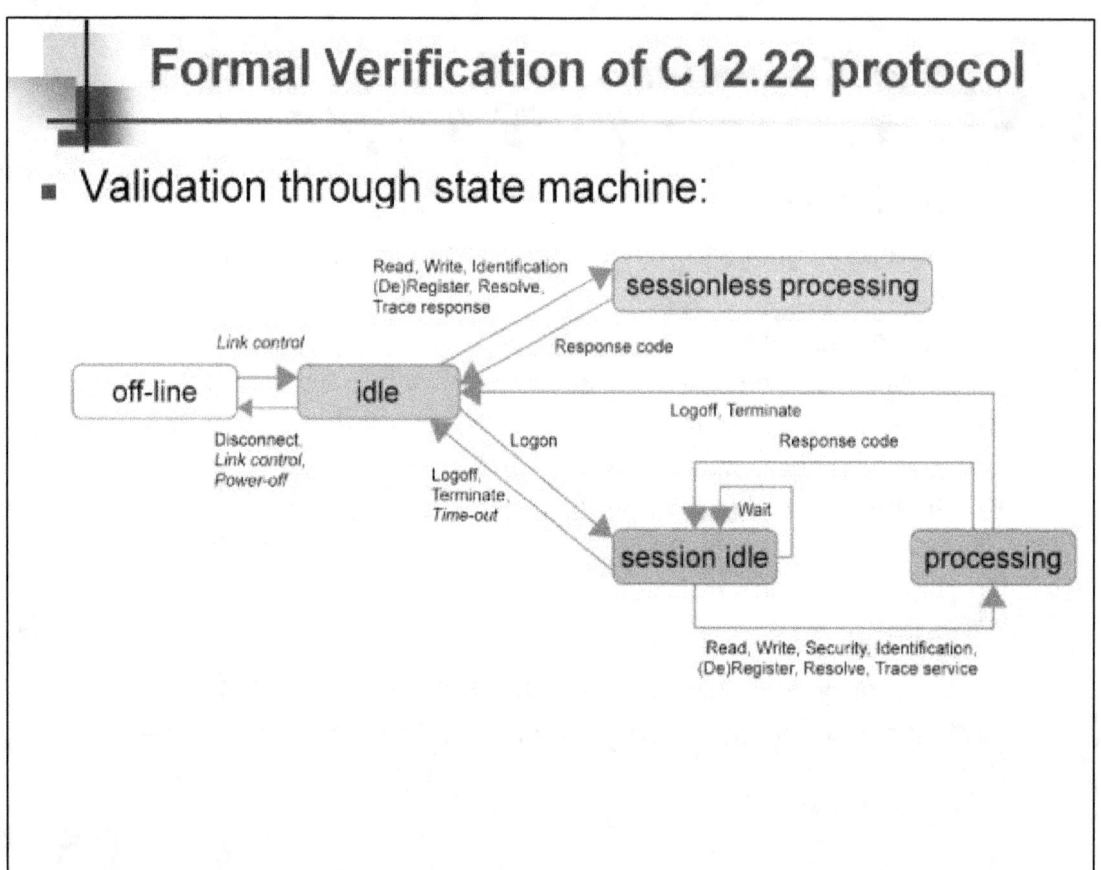

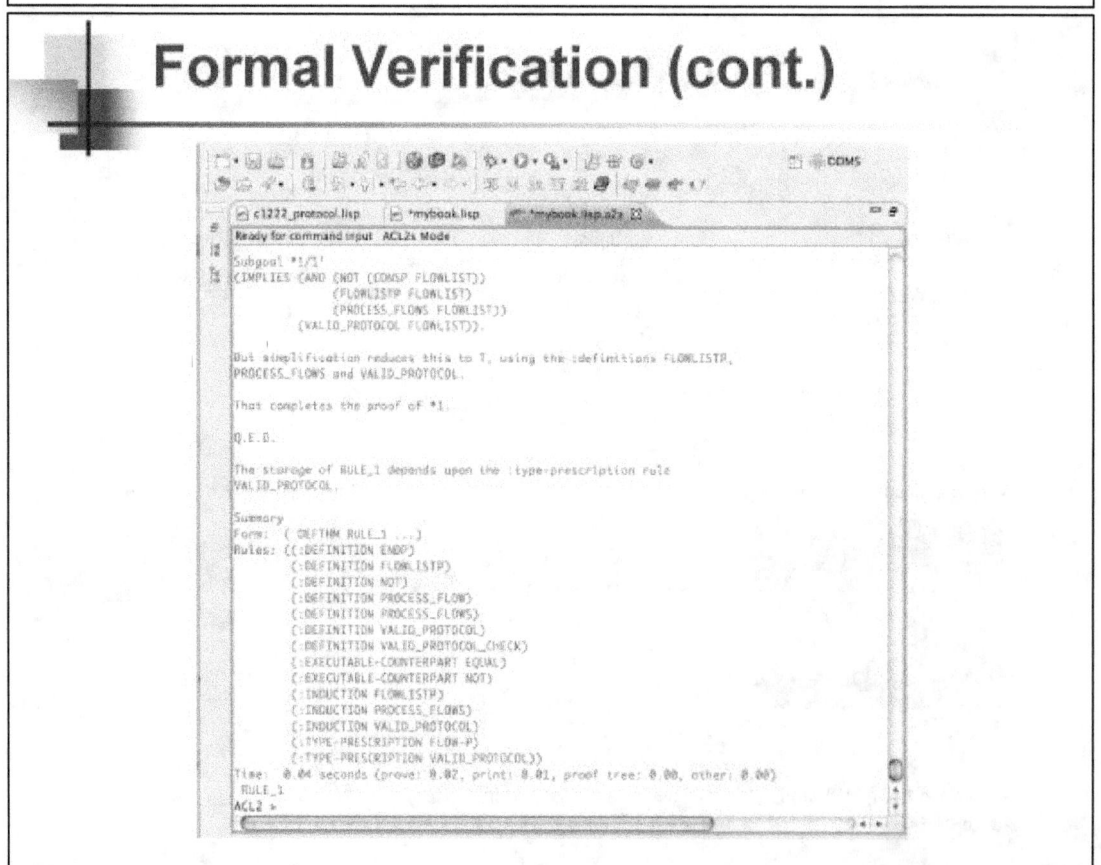

Attack Detection

- Violations at the network level

Type	Feature	Extracted automatically
Access	Origin/Dest.	From CE to meter
Data	Protocol	C12.22 over TCP/IP
Temporal	Frequency	1-2 per 1000 meters per day
Resource	Session size	< 100 bytes

- Violations at the application level

Type	Feature	Extracted automatically
Access	C12.19 tables	Table 0 (read), Table 3 (write)
Data	C12.19 values	Table 3, data: 0x01, offset: 0x00
Temporal	Session duration	< 1 minute
Resource	Services used	Logon, Full read, Partial write, Logoff

Defense Solutions (cont.)

System-aware detection and protection

Power-System Measurement Protection
and Bad-data Detection

Current Bad Data Detection Solutions: Residual-Based Approaches

- Need to account for possibility of bad data
  - **Bad data** definition from (*): "measurements that are grossly in error"
  - Bad data can potentially result in incorrect power-state estimates
- Measurement residuals – typical bad data detection for state estimation

  if $\|\mathbf{z} - \mathbf{H}\underline{x}\| \leq \tau$ *no bad measurements*

- Goal of residual approaches: detect corrupted power measurements

* A. Monticelli, State estimation in electric power systems: a generalized approach. Kluwer Academic Publishers, 1999.

Bad Data Detection: Residual -Based Approaches

- Coordinated attacks can work by creating "interacting bad-measurements" that satisfy the power flow solution equations, making them difficult or impossible to detect using conventional means

- *Residual-based approaches may be fundamentally insufficient against coordinated security compromises*

- One obvious approach:
  - Protect all measurements from compromises

System-Aware Measurement Protection

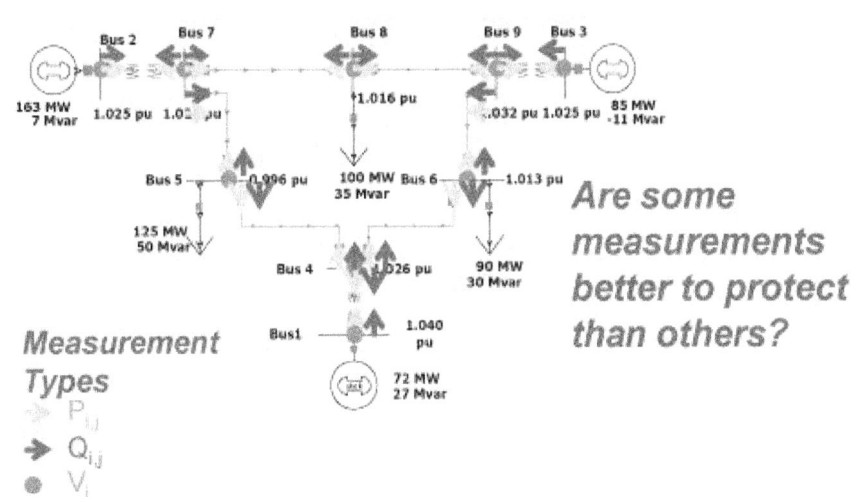

Measurement Types
- P$_{i,j}$
- Q$_{i,j}$
- V$_i$

Are some measurements better to protect than others?

System-Aware Measurement Protection

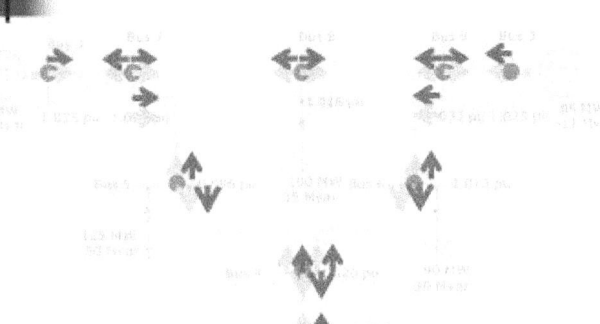

Measurement Types
- P$_{i,j}$
- Q$_{i,j}$
- V$_i$

We show that no attacks are possible if H'$_k$ has full rank

$$\begin{bmatrix} 0 \\ a_k \end{bmatrix} = \begin{bmatrix} H'' & H'_k \\ H_k' & H_{kk} \end{bmatrix} \begin{bmatrix} 0 \\ c_k \end{bmatrix}$$

$$0 = H'_k c_k$$
$$a_k = H_{kk} c_k$$

Accomplished by protecting *basic measurements*

Example: Basic Measurements

	i	j
P$_{ij}$	4	1
P$_{ij}$	2	7
P$_{ij}$	9	3
P$_{ij}$	5	4
P$_{ij}$	6	4
P$_{ij}$	7	5
P$_{ij}$	7	8
P$_{ij}$	8	9
Q$_{ij}$	4	1
Q$_{ij}$	8	9
Q$_{ij}$	7	2
Q$_{ij}$	3	9
Q$_{ij}$	4	5
Q$_{ij}$	4	6
Q$_{ij}$	5	7
Q$_{ij}$	8	7

Cost-Optimal Measurement Protection

- Protect a set of *Basic Measurements*[*]

  - it is necessary but not sufficient to protect n measurements, to detect stealthy false data injection attacks

  - it is necessary and sufficient to protect a set of basic measurements (BM) to detect stealthy false data injection attacks

  - approaches to identify BM already exist and well-studied

  - choices are available – the set of BM is not unique

  - each verifiable state variable (e.g., PMU) reduces number of measurements to be protected by one

  - approach validated on the IEEE 9,14,30,118, and 300 bus test systems

[*]R. B. Bobba, K. M. Rogers, Q. Wang, H. Khurana, K. Nahrstedt, T. J. Overbye, "Detecting False Data Injection Attacks on DC State Estimation," *First Workshop on Secure Control Systems (SCS 2010)*, April 2010.

Defense Solutions (cont.)

Integrated Cyber-Physical State Estimation

Cyber-Physical State Estimation (CPSE)*

- Co-utilize information from *cyber* and *power* network to (more precisely) determine the *state* of the *cyber-physical* system

Example

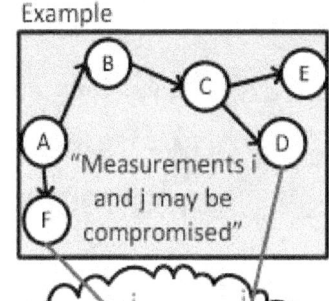

"Measurements i and j may be compromised"

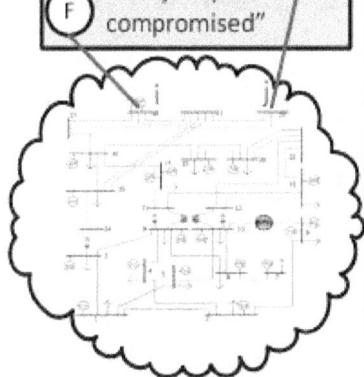

- Use combined ***information state*** to provide a scalable approach to detecting bad data caused by a cyber event

*S. A. Zonouz, K. M. Rogers, R. Berthier, R. B. Bobba, W. H. Sanders, T. J. Overbye, "CPIDS: A Cyber-Physical Intrusion Detection System for Power-Grid Critical Infrastructures," in review for *IEEE Transactions on Smart Grid*.

Algorithm Step 1: Potentially-bad Data Identification

- From IDS reports, we (probabilistically) know attacker's current privileges → From power network's topology, we know which measurements could/might have been modified by the adversary

Attack Graph

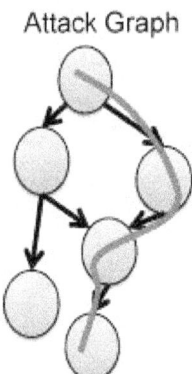

- Example:
  - network's topology
    - i-th measurement (by PMU_i): real power of the bus B2
  - IDS alerts
    - PMU_i is compromised
      → i-th measurement might have been corrupted!

34

Algorithm Step 2:
Power State Estimation & Verification

- Throw the potentially-bad data away, and run a power state estimation using the remaining power measurements

$$\mathbf{P_{ij}} = \mathbf{V_i^2}[-\mathbf{G_{ij}}] + \mathbf{V_i}\mathbf{V_j}[\mathbf{G_{ij}}\cos(\theta_i - \theta_j) + \mathbf{B_{ij}}\sin(\theta_i - \theta_j)]$$

$$\mathbf{Q_{ij}} = \mathbf{V_i^2}[-\mathbf{G_{ij}}] + \mathbf{V_i}\mathbf{V_j}[\mathbf{G_{ij}}\sin(\theta_i - \theta_j) + \mathbf{B_{ij}}\cos(\theta_i - \theta_j)]$$

- Compute $\| \mathbf{z} - \mathbf{H(\hat{x})} \|$, and identify the corrupted measurements
  - based on how much they differ from their estimates

35

CPSE Benefits

- Improved Bad-data Detection
  - Accuracy and Scalability
- Quick State Estimation Convergence
- Improved State Estimates

Defense Solutions (cont.)

System Contingency Analysis

Contingency Analysis (CA)

- Contingency analysis is a fundamental tool of power systems analysis
- Typically, a contingency analysis works with a power system model (power flow case) to determine potential problems
  - Full topology (node breaker) vs. planning models (bus branch)
- Answers the question: *"What happens when X goes out of service?"*

Contingency Analysis Results

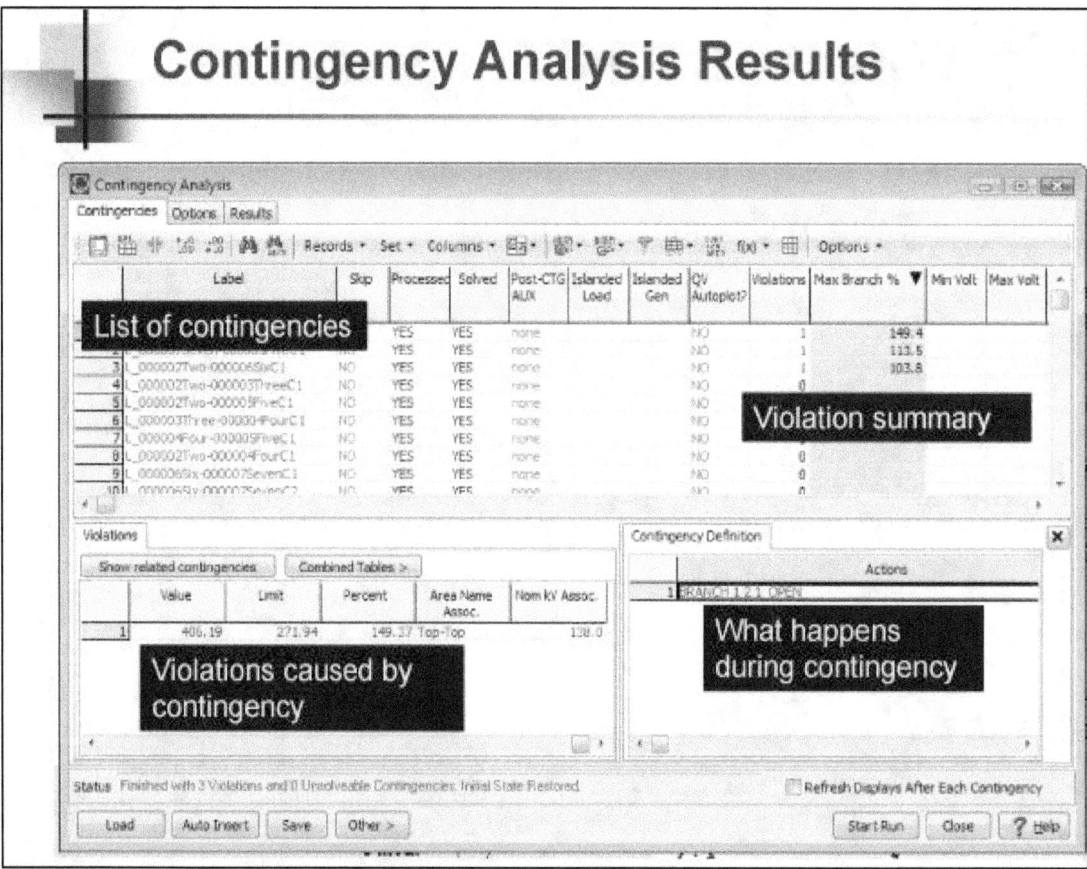

CA in Power System Operations

- State estimator runs every 2min or so
- After getting the state estimate real time contingency analysis (RTCA) runs on the estimated model
  - The list of contingencies must be picked carefully before being added to the RTCA contingency list
  - The RTCA list needs to include important contingencies, but it is time constrained

CA Solution Methods

- There are several ways of solving the contingency analysis
    - Full AC power flow (Slowest, Most accurate)
    - DC power flow (Fast, no voltage/var information)
    - Linear sensitivities (Fast, less sensitive to topology)
- There is the traditional engineering tradeoff between accuracy and speed
- All solution methods are used in practice

CA Solution Details

- Modeling a contingency accurately can be an intricate process
- The devil is in the details
- A few of the things that must be accounted for
    - Voltage controller and phase shifter response
    - AGC response
    - Special protection schemes / Breaker actions
    - Contingency modeling (full topology vs planning model)
- There is a lot that happens when a contingency is solved or even solving a power flow case

EMS and Planning Models

EMS Model
- Used for real-time operations
- Call this *Full-Topology* model
- Has node/breaker detail

Planning Model
- Used for off-line analysis
- We call this *Consolidated* model

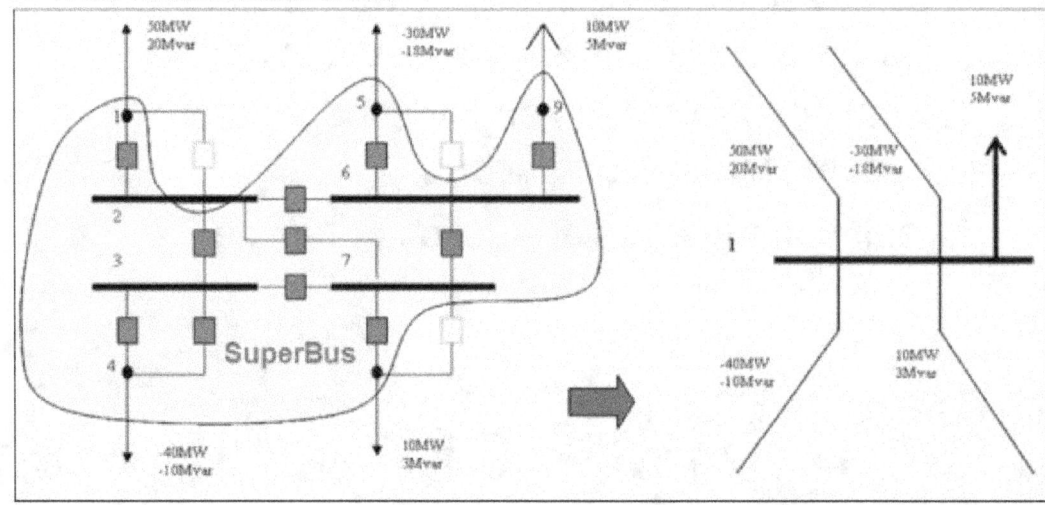

Traditional Contingency Analysis (CA)

- The "N-1" criteria is used to operate the system so that there will be no violations when any one element is taken offline

- Future requirements are strengthening the security criteria ("N-1-1") meaning many more contingencies need to be solved*

    - Once multiple outages begin to be considered, the size of the contingency list can grow very large
    - For 1000 lines
        - N-1 means solving 1000 line outages
        - N-2 means solving 499500 line outages (1000 choose 2)

*Charles Davis, Thomas Overbye: Linear Analysis of Multiple Outage Interaction. HICSS 2009: 1-8

Proposed System Contingency Analysis

- Question: "*What happens when X goes out of service?*"
  - X could be either a critical power component or cyber asset.
- Unlike traditional scenarios, cyber asset outages may be due to cyber adversaries

- Ongoing Research Topic!

Conclusions

- Criticality of cyber-physical infrastructure security:
  - Complex relationship between cyber and physical components
  - Importance of accurate state estimation → target of interest for adversaries:
    - Step-1: Cyber network exploits
    - Step-2: Physical system-aware attacks
- Requirements for advanced defense solutions:
  - Specification-based network intrusion detection tailored for cyber-physical system characteristics
  - System-aware measurement protection and bad-data detection
  - System-wide contingency analysis
- Contingency analysis as potential solution for a unified cyber-physical state estimation

Questions?

Robin Berthier   rgb@illinois.edu

Saman Zonouz   s.zonouz@miami.edu

14. False Data Injection Attacks in Smart Grid: Challenges and Solutions

Wei Yu
Department of Computer and Information Sciences
Towson University, Towson, MD 21252.
Email: wyu@towson.edu

Abstract—Smart Grid, as an energy-based Cyber-Physical System (CPS), is a new type of power grid that will provide reliable, secure, and efficient energy transmission and distribution. As the quality of assurance of monitoring data is essential to smart grid, in this talk we will first present two dangerous false data injection attacks, which target the state estimation and energy distribution in smart grid, respectively. We then present several defensive strategies against such attacks.

I. INTRODUCTION

The design of Cyber-Physical System (CPS) tends to integrate computing and communication capabilities with monitoring and control of entities in the physical world. Unlike traditional embedded systems, CPS is natural and engineered physical systems, which are integrated, monitored and controlled by an intelligent computational core [6]. A host of CPS, including the smart grid, process control systems, and transportation systems, are expected to be developed using advanced computing and communication technologies [4]. A smart grid is a typical energy-based CPS [1], which integrates a physical power transmission system with the cyber process of network computing and communication.

The quality of assurance of monitoring data is essential to smart grid. While most existing techniques for protecting power grids were designed to ensure system reliability (e.g., against random failures), recently there is a growing concern in smart grid initiatives on the protection against malicious cyber attacks. It was found that an adversary may launch attacks by compromising meters, hacking communication networks between meters and SCADA systems, breaking into the SCADA system through a control center office LAN, and breaking home area network and neighboring area network to compromise meters. Smart grid may operate in hostile environments and the sensor nodes lacking tamper-resistance hardware increases the possibility to be compromised by the adversary. Hence, the adversary can inject false measurement reports to disrupt the smart grid operation through the compromised meters and sensors. Those attacks are denoted as false data injection attacks and raise dangerous threats to the grid. In this following, we will first present two representative types of false data injection attacks and then discuss possible countermeasures.

II. FALSE DATA INJECTION ATTACKS

We now present two representative false data injection attacks, which target the state estimation and energy trans-

mission in smart grid.

(i) False Data Injection Attacks against State Estimation: It is critical for a smart grid to estimate its operating state based on meter measurements in the field and the configuration of grid. Recently, Liu *et al.* [3] developed a novel false data injection attack, which bypasses all the existing detection schemes and is therefore capable of arbitrarily manipulating power system states, posing dangerous threats to the control of power system. Differently, we considered the issue of how an adversary can choose the meters to compromise in order to cause the most significant deviation of the system state estimation [5]. We developed the least-effort attack model, which efficiently identifies the optimal set of meters to launch false data injection attacks for a fixed number of state variables. We also developed a heuristic algorithm to derive the results efficiently. The basic idea is listed below: the large power grid network is divided into a number of overlapping areas; the brute-force search method is used to identify the optimal set of meters for individual small areas and derive the optimal set of meters for the whole network. This heuristic algorithm was implemented on power system state manipulation using various IEEE standards buses (e.g., 9-bus, 14-bus, 30-bus, 118-bus, and 300-bus). Our data validated the feasibility and effectiveness of the developed scheme.

(ii) False Data Injection Attacks against Distributed Energy Distribution: Smart grid shall integrate the distributed energy resources and intelligently transmit energy to meet the requests from users. Hence, how to secure the distributed energy transmission and distribution process that utilizes the distributed energy resources and minimizes the energy transmission overhead is critical in smart grid. In our preliminary study, we studied the vulnerability of distributed energy transmission and distribution process and investigate novel false data injection attacks against distributed energy transmission and distribution process [2]. We considered several types of representative attacks, in which the adversary may manipulate the quantity of energy supply, the quantity of energy response, and link state of energy transmission. The forged data injected by those attacks will cause imbalanced demand and supply, increase the cost for energy distribution, disrupt the energy distribution causing some nodes energy outage in smart grid, and even manipulating energy price. Using graph and optimization theory, we formally modeled the attacks and quantitatively analyze their impact on energy distribution in smart grid. Our simulation data validated the effectiveness of those attacks

disrupting the effectiveness of energy distribution process, which may pose significant supplied energy loss, the increase of energy transmission cost, the number of outage users, and manipulation of energy price.

III. COUNTERMEASURES

To address those issues, we shall design the defensive countermeasures from the following perspectives: attack prevention, detection and response.

(i) Prevention: We shall enhance the network configuration to improve the resilience of grid to attacks. One way is to fully protect some of critical sensors and make them hard to be attacked. From the both attacks describe above, we can see that the false data injection attacks will become more difficult when we hide more system topology information. However, protecting all the sensors are impossible to realize in real-world practice because of deployment cost. Hence, we shall investigate the problem: given the limited number of sensors to be protected due to the cost constraint, how we can find the set of sensors to protect and make false data injection attacks difficult to deploy? We shall investigate the effectiveness of this countermeasure against the attacks when critical and redundant measurements are provided.

(ii) Detection: We shall develop robust intrusion detection techniques. Recall that in order to cause damage (e.g., manipulating the state estimation and energy transmission), the adversary needs to manipulate the sensor measurements. Obviously, if the adversary changes the true measurement value by a larger margin, he can manipulate smaller number of sensors, given a number of states to manipulate. In order to avoid from being detected by the standard anomaly detection, the adversary may become stealthy and tend to marginally change the sensor measurements, but still be able to manipulate the states to some extent. To address this problem, we shall analyze the properties of the false data injection attacks and find that the features with attacks always deviate much more from their means than measurements with random noises.

(iii) Response: Once an attack is detected, we shall develop schemes to localize the compromised devices and isolate the compromised devices from the grid. For example, to achieve this goal, one of schemes we shall consider is to adopt efficient watermarking-based forensic traceback scheme, which embeds secret signal (bits of 1 and 0) into the meter data stream. If the meter data stream is manipulated by any device during the transmission path, the receiver can correlate the received data stream with the secret signal and detect whether the data stream has been manipulated. By repeating the process over the transmission path, we can trace the origin which manipulates the data.

IV. CONCLUSION

In this talk, we first present two types of false data injection attacks against smart grid operation. One is to disrupt the state estimation of smart grid and the other is to disrupt the energy distribution of smart grid. We then present several possible countermeasures, including attack prevention, detection and response.

REFERENCES

[1] Nsf workshop on new research directions for future cyber-physical energy systems. Technical report, http://www.ece.cmu.edu/ nsf-cps/, Baltimore, MD, 2009.

[2] J. Lin, W. Yu, X. Yang, G. Xu, and W. Zhao. On false data injection attacks against distributed energy routing in smart grid. In *Proc. of ACM/IEEE Third International Conference on Cyber-Physical Systems (ICCPS) (part of CPS Week 2012)*, April 2012.

[3] Y. Liu, M. K. Reiter, and P. Ning. False data injection attacks against state estimation in electric power grids. In *Proceedings of the 16th ACM conference on Computer and communications security*, November 2009.

[4] J. Wang, D. Li, Y. Tu, P. Zhang, and F. Li. A survey of cyber physical systems. In *Proc. of IEEE International Conference on Cyber Technology in Automation, Control, and Intelligent Systems*, March 2011.

[5] Q. Yang, J. Yang, W. Yu, N. Zhang, and W. Zhao. On a hierarchical false data injection attack of power system state estimation. In *Proc. of IEEE Globe Communication (Globecom)*, December 2011.

[6] T. Znati. Security for emerging cyber-physical systems research challenges and directions. In *Proc. of First International Workshop on Data Security and PrivAcyin wireless Networks Panel*, July 2010.

False Data Injection Attacks in Smart Grid: Challenges and Solutions

Dr. Wei Yu
Assistant Professor
Department of Computer & Information Sciences
Towson University
http://www.towson.edu/~wyu
Email: wyu@towson.edu

NIST Cyber Security for CPS Workshop   Towson University   Wei Yu

Research Projects

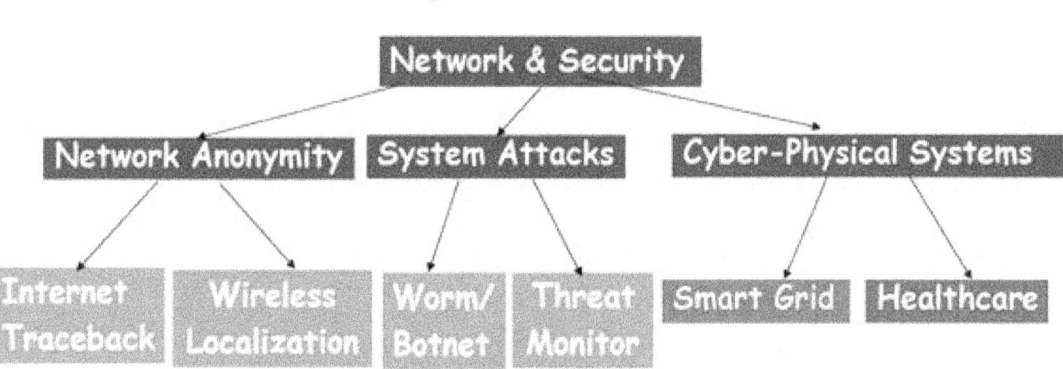

1. Qinyu Yang, Jie Yang, Wei Yu, Nan Zhang, and Wei Zhao, "False Data Injection Attack Against Power System State Estimation: Modeling and Defense", in Proceedings of IEEE Globecom 2011 (journal version is under submission to IEEE TPDS)
2 Jie Lin, Wei Yu, Guobin Xu, Xinyu Yang and Wei Zhao, "On False Data Injection Attacks against Distributed Energy Routing in Smart Grid," in Proceedings of IEEE/ACM International Conference on Cyber Physical System (ICCPS), 2012.
3. Xinyu Yang, Jin Lin, Paul Moulema, Wei Yu, Xinwen Fu, and Wei Zhao, "A Novel En-route Filtering Scheme against False Data Injection Attacks in Cyber-Physical Networked Systems," in Proceedings of IEEE International Conference on Distributed Computing Systems (ICDCS), 2012.

http://www.towson.edu/~wyu

NIST Cyber Security for CPS Workshop   Towson University   Wei Yu

Outline

☐ Overview

☐ False Data Injection Attack against Grid System State Estimation

☐ False Data Injection Attack against Energy Distribution

☐ Final Remarks

NIST Cyber Security for CPS Workshop   Towson University   Wei Yu

Traditional Grid

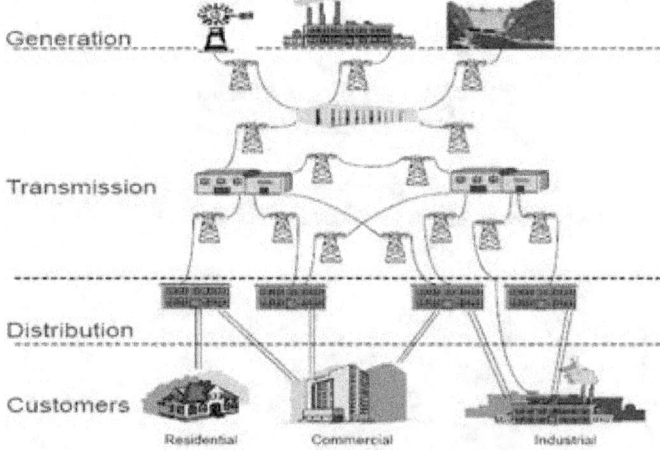

☐ Centralized one way electricity delivery from generation to end-users
☐ Over-provision energy generation and load control
☐ Limited automation and situational awareness
☐ Lack of customer-side management

NIST Cyber Security for CPS Workshop   Towson University   Wei Yu

Smart Grid: An Energy-based Internet

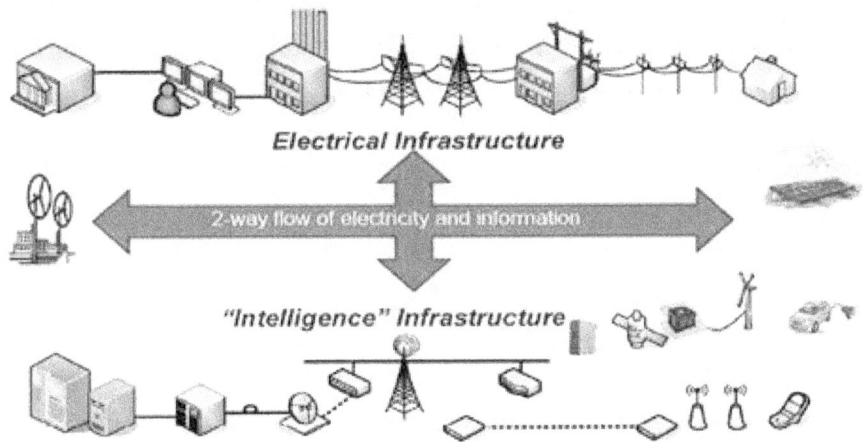

Electrical Infrastructure

2-way flow of electricity and information

"Intelligence" Infrastructure

❑ Smart Grid will comprise a vast array of devices and systems with two-way communication and control capabilities

❑ An energy-based Internet

NIST Cyber Security for CPS Workshop   Towson University   Wei Yu

Smart Grid as an Energy-based Cyber-Physical System (CPS)

❑ Cyber – computation, communication, and control that are discrete, logical, and switched

❑ Physical – natural and human-made systems governed by the laws of physics and operating in continuous time

❑ Cyber-Physical Systems – systems in which the cyber and physical systems are tightly integrated at all scales and levels

❑ Smart grid is a typical CPS, which integrates a physical power transmission system with the cyber process of network computing and communication

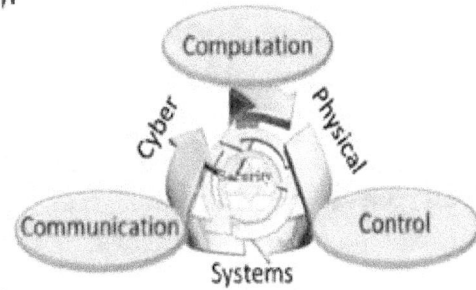

NIST Cyber Security for CPS W   Wei Yu

245

Key Services in Smart Grid (NIST)

- **Energy distribution management:** Making the energy distribution system more intelligent, reliable, self-repairing, and self-optimizing
- **Distributed renewable energy integration:** Integrating distributed renewable-energy generation facilities, including the use of renewable resources (i.e., wind, solar, thermal power, and others)
- **Distributed energy storage:** Enabling new storage capabilities of energy in a distributed fashion, and mechanisms for feeding energy back into the energy distribution system
- **Electric vehicles-to-grid:** Enabling large-scale integration of plug-in electric vehicles (PEVs) into the transportation system
- **Grid monitoring and management:** Enabling the demand response and consumer energy efficiency
- **Smart metering infrastructure:** Providing customers real-time (or near real-time) pricing of electricity and can help utilities achieve necessary load reductions

NIST Cyber Security for CPS Workshop   Towson University   Wei Yu

Real-World Cyber Attacks in Smart Grid

- Cybercriminals compromise computers anywhere they can find them (even in smart grid systems)
  - January 2003, computers infected by the Slammer worm shut down safety display systems at power plant in Ohio
- Disgruntled employees can be the major source of targeted computer attacks against systems
  - Contractor launches an attack on a sewage control system in Queensland in 2000
  - More than 750,000 gallons of untreated sewage released into parks, rivers, and hotel grounds
- Terrorists, activists, and organized criminal groups
  - In 2008, there was evidence of computer intrusions into some European power utilities
  - In 2010, Stuxnet worm provides a blueprint for aggressive attacks on control systems

NIST Cyber Security for CPS Workshop   Towson University   Wei Yu

False Data Injection Attacks

- ❑ Smart grid may operate in hostile environments
- ❑ Meters and sensors lacking tamper-resistance hardware increases the possibility to be compromised
- ❑ The adversary may inject false measurement reports to the disrupt the smart grid operation through the compromised meters and sensors
- ❑ Those attacks denoted as false data injection attacks
  - ○ It can disrupt the grid system state estimation
  - ○ It can disrupt the energy distribution

NIST Cyber Security for CPS Workshop   Towson University   Wei Yu

Outline

- ❑ Overview
- ❑ False Data Injection Attack against Grid System State Estimation
- ❑ False Data Injection Attack against Energy Distribution
- ❑ Final Remarks

NIST Cyber Security for CPS Workshop   Towson University   Wei Yu

Objectives

- ☐ Smart grid shall provide reliable, secure, and efficient energy transmission and distribution
- ☐ State estimation is a very critical component in power grid system operation
  - ○ Used by Energy Management Systems (EMS) at the control center to ensure that the power grid is in the desired operation states
- ☐ Objectives of this research
  - ○ Modeling the false data injection attacks against power system state estimation
  - ○ Studying countermeasures against such attacks

Power System Operation

- ☐ The operation condition of a power grid over time can be determined if the network model and voltages at every system bus are known.

- ☐ State estimator (SE) uses Supervisory Control and Data Acquisition (SCADA) data and system model to estimate the system states (e.g., voltages at all system buses) in real time.

State Estimation Process

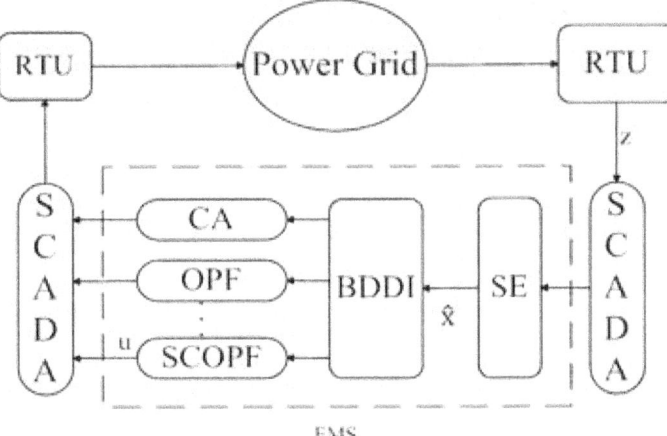

EMS: Energy management system
RTU: Remote terminal unit
BDDI: Bad data detection and identification
CA: Contingency analysis
OPF: Optimal power flow
SCOPF: Security constrained OPF

Algorithm for State Estimation

❑ The state estimation can be formalized by

$$\mathbf{z} = h(\mathbf{x}) + \mathbf{e}$$

z: Measurement vector (bus voltages, bus active an reactive power flows, and branch active and reactive power flows)

x: State vector (bus voltage magnitudes & phase angles)

h(x): Nonlinear vector function determined by the system topology

e: Error vector, cov(e)=R

❑ Most existing state estimators use a weighted least squares (WLS) method to minimize the objective error function

$$\min_{\mathbf{x}}: \mathrm{J}(\mathbf{x}) = [\mathbf{z} - h(\hat{\mathbf{x}})]^T \mathbf{R}^{-1} [\mathbf{z} - h(\hat{\mathbf{x}})]$$

Bad Data Detection and Identification

❏ What is bad data?
- ○ Random errors can be filtered by the state estimator
- ○ Large measurement errors occur when meters have biases, drifts or wrong connections

❏ How to deal with bad data?
- ○ Detection and identification of bad data are done only after the estimation process by processing the measurement residuals
- ○ Largest normalized residual (LNR) test: the presence of bad data is determined by a hypothesis test if

$$J(x) = \|z - H\hat{x}\|^2_{R^{-1}} \geq \tau.$$

NIST Cyber Security for CPS Workshop   Towson University   Wei Yu

---

False data Injection Attacks

❏ Liu et al., "False data injection attacks against state estimation in electric power grids," in Proceedings of ACM Computer Communication Security (CCS), November 2009

❏ By taking advantage of the configuration information of a power system, the adversary can inject malicious measurements
- ○ Mislead the state estimation process without being detected by existing bad data detection techniques.

$$z_a = z + a, \hat{x}_{bad} = \hat{x} + c$$

$$\|z_a - H\hat{x}_{bad}\| = \|z + a - H(\hat{x} + c)\|$$
$$= \|z - H\hat{x} + (a - Hc)\|$$
$$= \|z - H\hat{x}\|$$
$$when \quad a = Hc$$

NIST Cyber Security for CPS Workshop   Towson University   Wei Yu

False data Injection Attacks

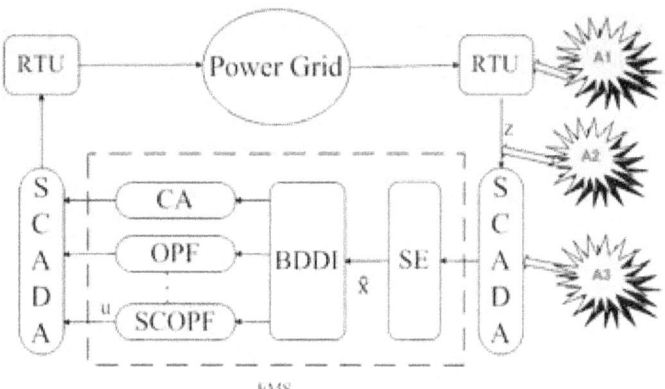

□ Assumptions

- ○ The adversary has an accurate model of the power system
- ○ The adversary knows the state estimation and bad data detection methods
- ○ The adversary will compromise as few meters as possible

Our Contributions

□ When the attackers are constrained to inject false data into specific number of state variables, what is the least number of meters should they compromise?

- ○ We develop a least-effort attack model to identify the optimal set of meters to launch false data injection attacks.
- ○ We show that the problem can be reduced to a NP-hard problem - minimum subadditive join problem.
- ○ We develop a heuristic algorithm to derive the results efficiently.
- ○ We develop countermeasures to defend against such attacks.

Hierarchical Approach

☐ We first divide the large-scale power system into N overlapping areas, find the suboptimal sets of sensor measurements in each area.

☐ We then can obtain an optimal solution for the whole system.

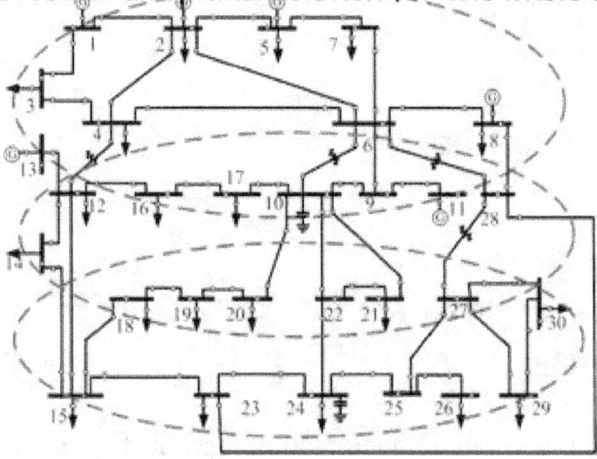

Example of IEEE 30-bus with Measurements

Performance of Brute-force Search

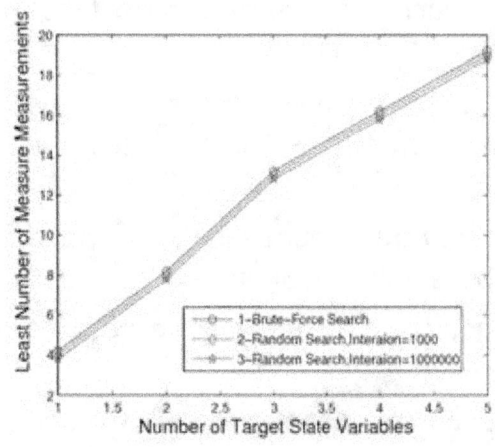

Brute-force Search for IEEE 9-bus

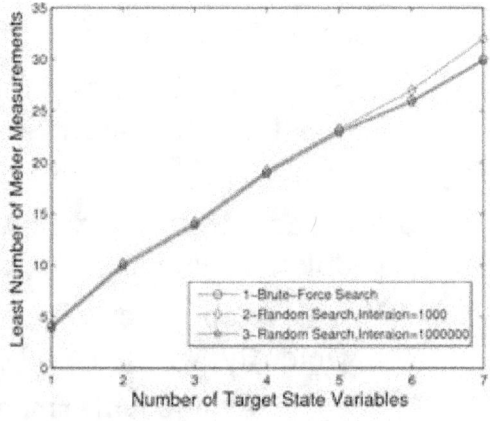

Brute-force Search for IEEE 14-bus

Performance of Hierarchical Search

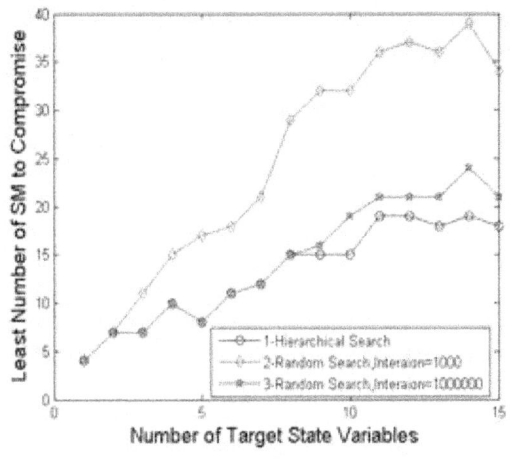

Hierarchical Search for IEEE 30-bus

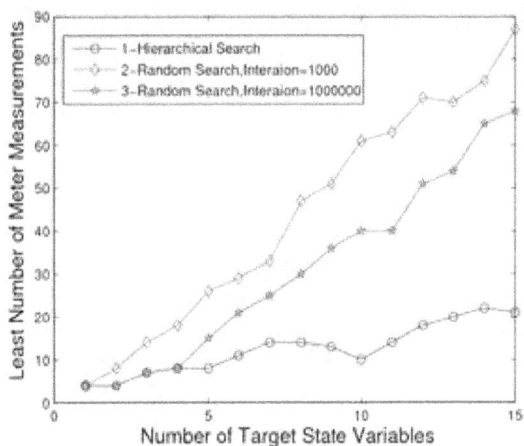

Hierarchical Search for IEEE 118-bus

NIST Cyber Security for CPS Workshop   Towson University   Wei Yu

Performance of Hierarchical Search

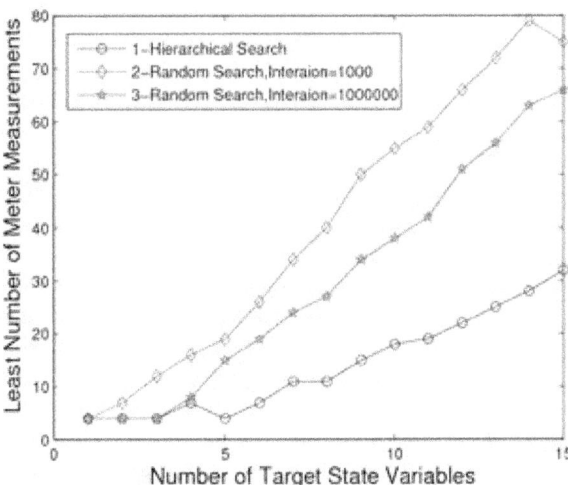

Hierarchical Search for IEEE 300-bus

NIST Cyber Security for CPS Workshop   Towson University   Wei Yu

Countermeasures

❏ System Protection
- Some of the measurement play a critical role in determining a specific state variable, while others are redundant to improve the accuracy of state estimation.
- How to select a set of sensors to protect and make attacks difficult to deploy.

❏ Anomaly Detection
- Spatial-based detection
  - Treat all the measurements received at a certain time as a unity and the accumulated deviation of all compromised measurements will be significant.
- Temporal-based detection
  - Consider the fact that the adversary needs to manipulate sensor measurements over time
  - Develop the nonparametric cumulative sum (cusum) change detection technique.

NIST Cyber Security for CPS Workshop   Towson University   Wei Yu

Preliminary Evaluation Results

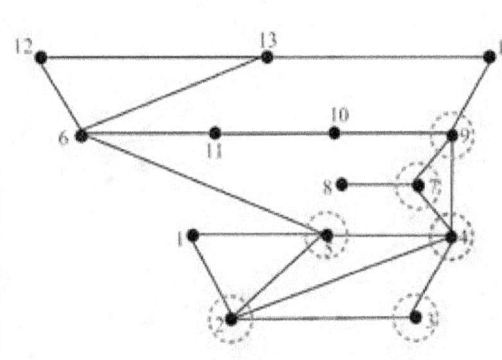

Topology of IEEE 14-bus System

NIST Cyber Security for CPS Workshop   Towson University   Wei Yu

Ongoing Research

☐ Attacks in dynamic state estimation
- ○ The dynamic state estimation can obtain complete, coherent, and real-time dynamic states.
- ○ We investigate attack schemes against dynamic state estimation and countermeasures.

☐ Attacks against control algorithms
- ○ Applications such as contingency analysis, optimal power flow, and economic dispatch can be the target.
- ○ Attacks will make the control center generate false control signals.

Outline

☐ Overview

☐ False Data Injection Attack against Grid System State Estimation

☐ False Data Injection Attack against Energy Distribution

☐ Final Remarks

Objectives

❏ Smart grid shall provide reliable, secure, and efficient energy transmission and distribution
  ○ Efficiently utilize the distributed energy resources
  ○ Minimize the energy transmission overhead
❏ Objectives of this research
  ○ Study the vulnerability of distributed energy routing process
  ○ Investigate false data injection attacks against the energy routing process

NIST Cyber Security for CPS Workshop   Towson University   Wei Yu

Smart Meters

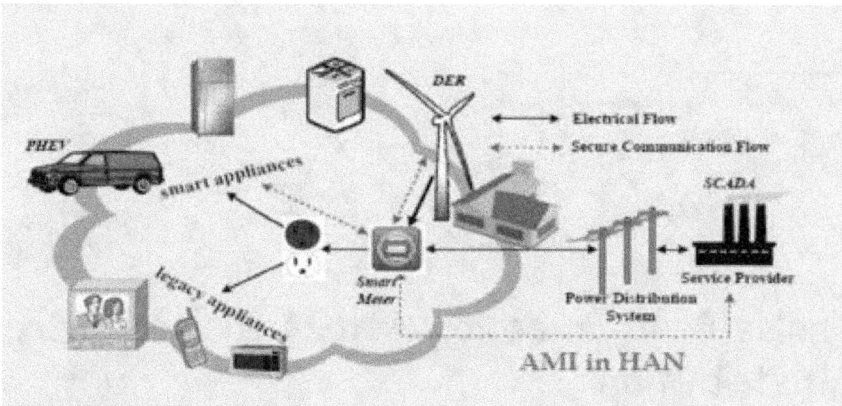

❏ Smart meter computes consumption and sends the information to utility for monitoring and billing purpose.
❏ Smart meter has the ability to disconnect-reconnect remotely and control the user appliances and device to manage load and demands.
  ○ Examples: reduce bill for customer & optimize power flow for utility

NIST Cyber Security for CPS Workshop   Towson University   Wei Yu

Attacks against Smart Meters

- Smart meter is "computer" and all cyber attacks can be applied
- Widespread use of smart meters
- A potentially large number of opportunities for the adversary
  - Forging the demand request of a smart meter (e.g., requesting a large amount of energy).
  - Misleading the electric utility into making incorrect decision about local or regional usage and capacity.
  - Nightmare scenario: deployed millions of smart meters and controlled by adversary
    - Interrupt the supply/demand process and cause disastrous consequences

NIST Cyber Security for CPS Workshop   Towson University   Wei Yu

Network Model

- The input energy of demand-nodes should be equal to their demanded energy.
- The output energy of supply-nodes should be less than energy that they could provide to the grid.
- The energy transmitted on a link should be less than the link capacity.

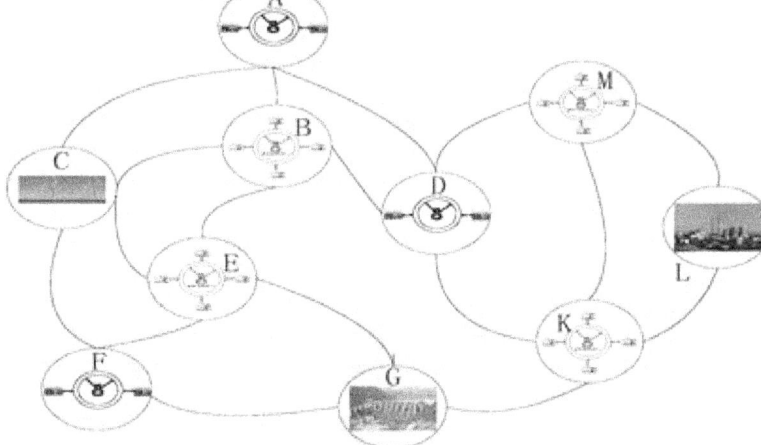

NIST Cyber Security for CPS Workshop   Towson University   Wei Yu

Distributed Energy Management

☐ The formalization of distributed energy management is

Objective. $\quad Min\left(Cost = \dfrac{1}{2}\sum_{l_{ij}\in L} Cost_{ij}\cdot E_{ij}\right)$

$$S.t.\begin{cases} \forall v\in N_P \quad \sum_{i\in N_v} E_{vi}\leq P_v \\[2ex] \forall u\in N_D \quad \sum_{j\in N_u} E_{uj}=-D_u \\[2ex] \forall l_{ij}\in L \quad E_{ij}=-E_{ji} \\[2ex] \forall l_{ij}\in L \quad \left|E_{ij}\right|\leq Load_{ij} \end{cases}$$

E_{ij} is the energy transmitted on link L_{ij};
N_P is the supply-nodes set;
N_D is the demand-nodes set;
P_v is the residual energy of node v;
D_u is demanded energy of node u.
$Load_{ij}$ is the link capacity of link L_{ij}

NIST Cyber Security for CPS Workshop   Towson University   Wei Yu

Example

Objective. $\quad Min\left\{ Cost(\dfrac{1}{2}\cdot\sum_{E_{IJ}}(|E_{IJ}|\cdot Cost_{IJ}))\right\}$

S.t.
$$\begin{cases} E_{AC}+E_{AB}+E_{AD}=-D_A \\ E_{DA}+E_{DB}+E_{DK}+E_{DM}=-D_D \\ E_{FC}+E_{FE}+E_{FG}=-D_F \\ E_{CE}+E_{CA}+E_{CF}+E_{CB}\leq P_C \\ E_{BC}+E_{BA}+E_{BE}+E_{BD}\leq P_B \\ E_{EC}+E_{EB}+E_{EF}+E_{EG}\leq P_E \\ E_{MD}+E_{ML}+E_{MK}\leq P_M \\ E_{LM}+E_{LK}\leq P_L \\ E_{KM}+E_{KD}+E_{KL}+E_{KG}\leq P_K \\ \forall E_{IJ}, \quad E_{IJ}+E_{JI}=0 \\ \forall E_{IJ}, \quad 0\leq|E_{IJ}|\leq Load_{IJ} \end{cases}$$

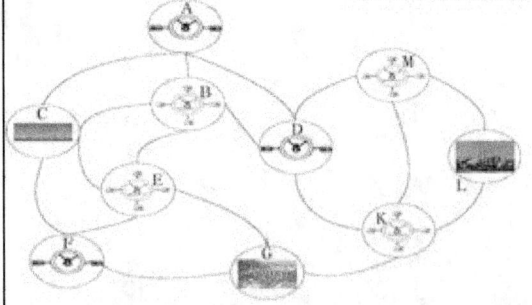

A, D, and F are demand nodes
Others are supply nodes

Towson University   Wei Yu

False data Injection Attacks

❑ Injecting False Energy Data

- ○ Energy-request Deceiving Attack
  - · The adversary compromises demand-nodes and injects forged quantity of demanded energy.
- ○ Energy-supply Deceiving Attack
  - · The adversary compromises supply-nodes and injects forged quantity of energy that the supply-nodes could provide to the grid.

❑ Injecting False Link-state Data

- ○ Claiming invalid energy links as valid
- ○ Claiming valid energy links as invalid

Metrics

❑ Supplied energy loss
  - ○ Energy loss due to forged energy data from energy supply perspective

❑ Energy transmission cost
  - ○ The increased total energy transmission cost caused by forged energy data

❑ The number of outage users
  - ○ Some users could be outage due to the unbalance energy distribution caused by attacks

Energy-request Deceiving Attack

❑ In this scenario, the formalization of compromised distributed energy management is

$$Objective. \quad Min\left(Cost^* = \frac{1}{2}\sum_{l_{ij} \in L} Cost_{ij} \cdot E_{ij}\right)$$

$S.t.$

$$
\begin{cases}
\forall v \in N_P & \sum_{i \in N_v} E_{vi} \leq P_v \\
\forall u \in N_D & \sum_{j \in N_u} E_{uj} = -D_u \\
\forall u^* \in N_{D^*} & \sum_{j \in N_u} E_{u^*j} = -D_{u^*}^* \leq T_E \\
\forall l_{ij} \in L & E_{ij} = -E_{ji} \\
\forall l_{ij} \in L & |E_{ij}| \leq Load_{ij}
\end{cases}
$$

u* is the compromised demand-nodes;
D*_u* is the forged demanded energy;
T_E is the threshold

NIST Cyber Security for CPS Workshop   Towson University   Wei Yu

Energy-request Deceiving Attack (cont.)

❑ **Supplied Energy Loss:**

$$\Delta D^n = \sum_{u_i \in N_{D^*}} D_{u_i}^* - D_{u_i}$$

When the grid has enough energy, the forged demanded energy will be provided by supply-nodes, and then the supplied energy loss would occur.

NIST Cyber Security for CPS Workshop   Towson University   Wei Yu

Energy-request Deceiving Attack (cont.)

☐ **Energy Transmission Cost**:

$$\Delta Cost_n = Min(Cost_n^*) - Min(Cost)$$

As the analysis in our paper, with the increase of forged demanded energy D_u^* , the energy transmitted on links would be increase, and we can always have $\Delta Cost_n > 0$. Hence, energy-request deceiving attack can certainly increase the energy transmission cost.

Energy-request Deceiving Attack (cont.)

☐ **The number of outage users:**

With the objective of minimize the number of outage demand-nodes, the problem can be represented by

$$Objective. \quad s = Min\left(\| N_D^{'} \|\right)$$

$$S.t.$$

$$\sum_{u \in N_D^{'}} D_u \geq \sum_{u \in N_D} D_u - \sum_{v \in N_P} P_v$$

$N_D^{'}$ is the set of outage users.

Energy-supply Deceiving Attack

❏ In this scenarios, the formalization of compromised distributed energy management is

$$Objective. \quad Min\left(Cost^* = \frac{1}{2} \sum_{l_{ij} \in L} Cost_{ij} \cdot E_{ij} \right)$$

$S.t.$

$$\begin{cases} \forall v \in N_P & \sum_{i \in N_v} E_{vi} \le P_v \\ \forall v^* \in N_{P^*} & \sum_{i \in N_{v^*}} E_{v^*i} \le P_{v^*}^* \\ \forall u \in N_D & \sum_{j \in N_D} E_{uj} = -D_u \\ \forall l_{ij} \in L & E_{ij} = -E_{ji} \\ \forall l_{ij} \in L & |E_{ij}| \le Load_{ij} \end{cases}$$

v^* is the compromised supply-nodes;

$P_{v^*}^*$ is the forged energy that supply-node could provide to the grid.

NIST Cyber Security for CPS Workshop   Towson University   Wei Yu

Energy-supply Deceiving Attack

❏ Claiming more energy than supply-node can provide
  ○ Demand-node cannot obtain expected energy
❏ Claiming less energy than supply-node can provide
  ○ Increase energy transmission cost
  ○ Increase number of outage users

NIST Cyber Security for CPS Workshop   Towson University   Wei Yu

Injecting False Link-state Data

- ☐ Claiming invalid energy links as valid
  - ○ Demand node cannot obtain enough requested energy
  - ○ Disrupt energy transmission in the grid

- ☐ Claiming valid energy links as invalid
  - ○ Small number of links compromised—total transmission cost increase
  - ○ Large number of links compromised—total transmission cost decrease

Performance Evaluation

- ☐ **Topology**: The simplified version of the US smart grid.
- ☐ **Data set**: 2009 US Energy Information Administration State Electricity Profiles.
- ☐ **Length of the energy links**: Computed using Google map.
- ☐ **Metrics**: Increased transmission cost, User outage rate, and Supplied energy loss.

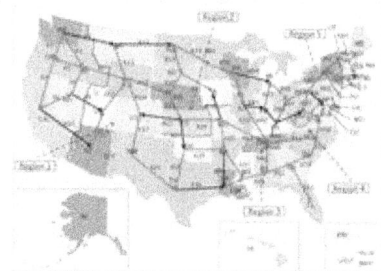

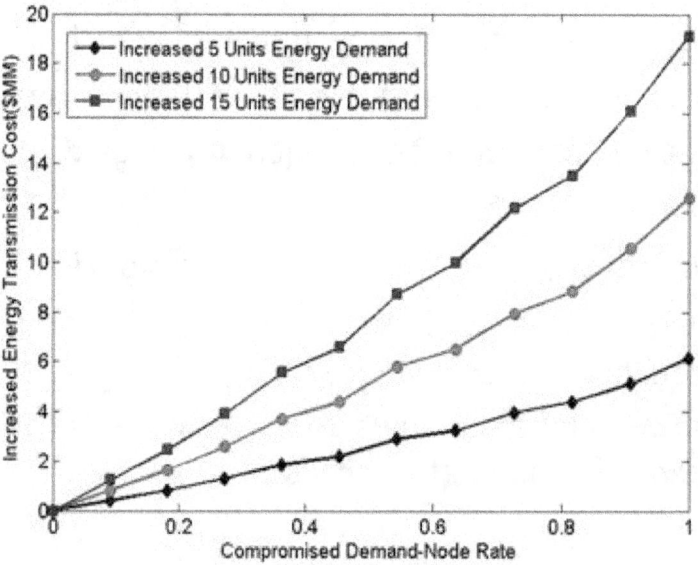

Fig. 3 Increased Energy Cost vs. Compromised Demand-Node Rate

NIST Cyber Security for CPS Workshop   Towson University   Wei Yu

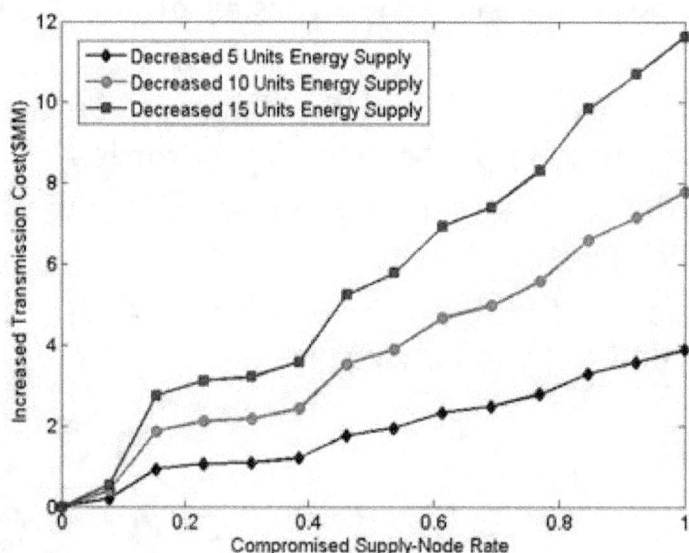

Fig. 4 Increased Energy Transmission Cost vs. Compromised Supply-Node Rate

NIST Cyber Security for CPS Workshop   Towson University   Wei Yu

Performance Evaluation (cont.)

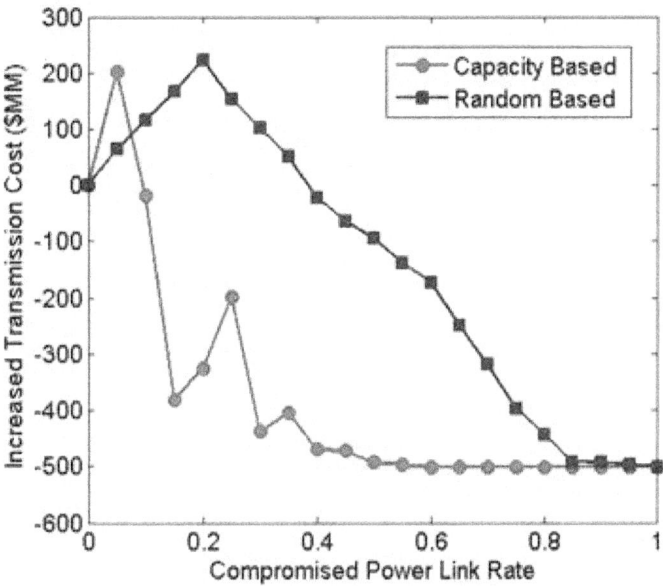

Fig. 5 Energy Transmission Cost vs. Compromised Energy Link Rate

NIST Cyber Security for CPS Workshop   Towson University   Wei Yu

Performance Evaluation (cont.)

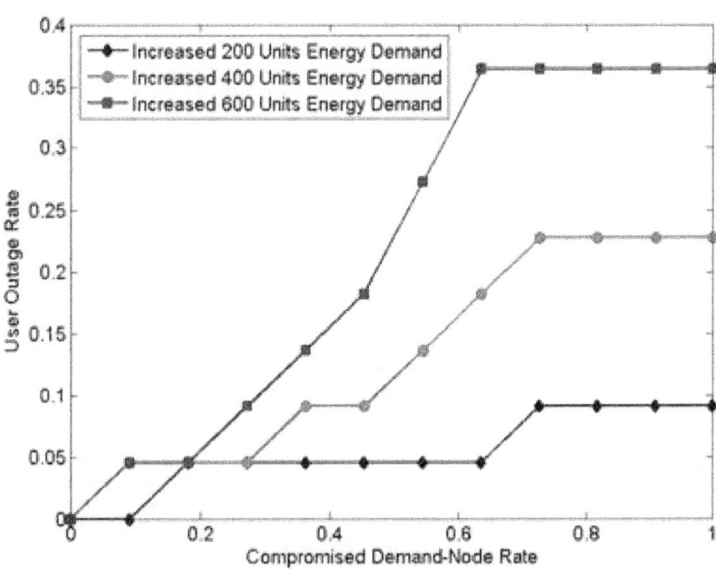

Fig. 6 User Outage Ratio vs. Compromised Demand-Node Rate

NIST Cyber Security for CPS Workshop   Towson University   Wei Yu

265

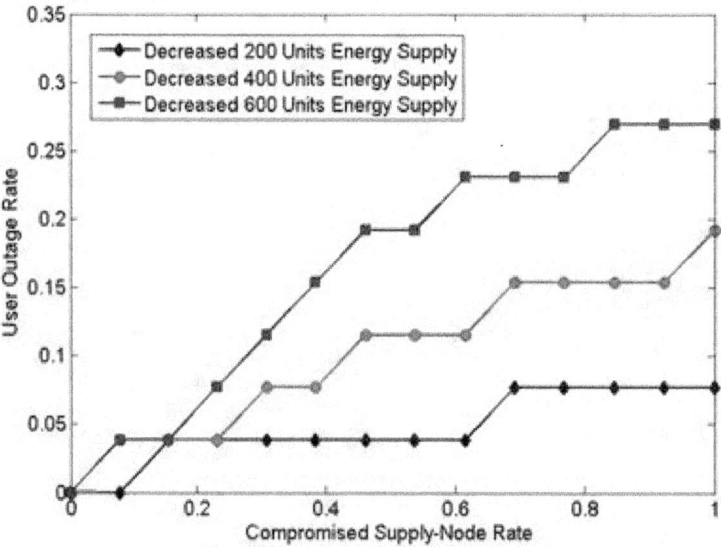

Fig. 7 User Outage Rate vs. Compromised Supply-Node Rate

NIST Cyber Security for CPS Workshop   Towson University   Wei Yu

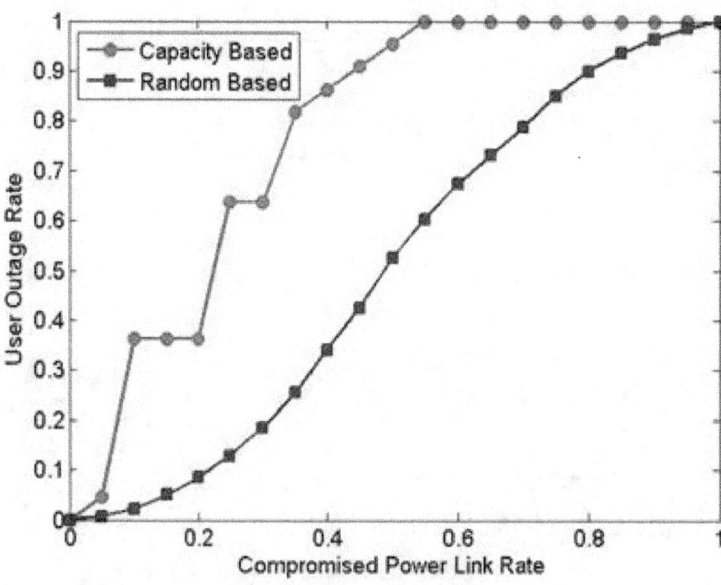

Fig. 8 User Outage Rate vs. Compromised Energy Link Rate

NIST Cyber Security for CPS Workshop   Towson University   Wei Yu

Performance Evaluation (cont.)

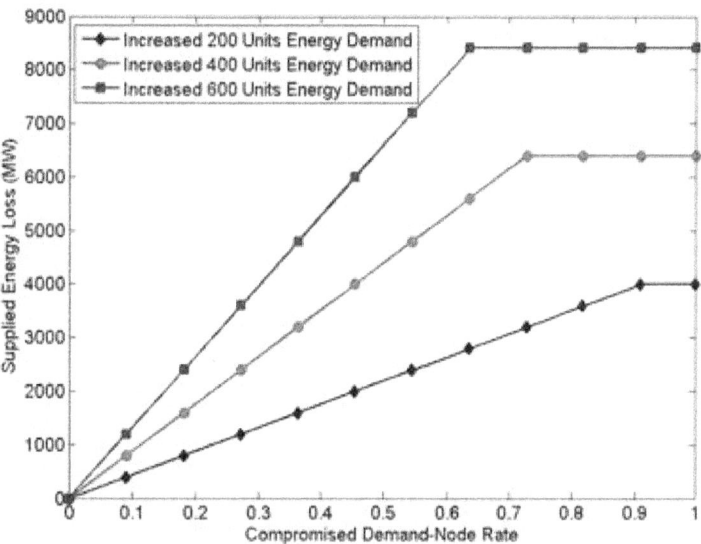

Fig. 9 Supplied Energy Loss vs. Compromised Demand-Node Rate

NIST Cyber Security for CPS Workshop   Towson University   Wei Yu

Final Remarks

❑ False data injection attacks against power system state estimation
  ○ Modeling attacks
  ○ Developing countermeasures

❑ False data injection attacks against energy routing process
  ○ Exploring the space of attack strategies
  ○ Modeling and analysis

❑ Ongoing research
  ○ Explore other attacks (data integrity, timing, and others)
  ○ Defend against those attacks
    • Prevention, detection and response

NIST Cyber Security for CPS Workshop   Towson University   Wei Yu

Thank You!

Questions?

NIST Cyber Security for CPS Workshop   Towson University   Wei Yu

15. Conclusion

The goals of the workshop were to look at recent (2 – 3 years) research results and deployment experiences that have occurred in cyber-physical areas across multiple industries. (e.g., healthcare, manufacturing, automotive, electric smart grid), and to determine if there are security requirements that are unique to CPS as opposed to strictly cyber or physical systems. Through these presentations on recent CPS cybersecurity research ideas and themes emerged. While some of these ideas are more critical from a security and safety technical standpoint, those that are policy and business oriented are equally crucial to implementing adequate security within CPS.

Attendees heard about the difficulty of detecting attacks on CPS. First, being able to tell the difference between an attack and a system failure can be difficult. Detecting an attack by analyzing massive amounts of data is both time consuming and difficult, as well as costly. In addition to attacks from outsiders, there are threats from inside sources. Finding malicious code inserted by an insider in over 100 million lines of code can be virtually impossible. All of these things make detecting cyber attacks challenging in CPS.

Many presenters stressed the need for improved resiliency. While attacks to CPS will happen, a greater question is how will a system perform during and after an attack? Will the CPS continue to function at all? What will the consequences be? Without knowing the answers to these questions, building in layers of security and improving resiliency are critical to the continuing operations of CPS, especially when lives are dependent upon this continued operation. While a need for improved resiliency is not unique to CPS, the consequences of a system failure can be greater (potential loss of life) than in strictly cyber systems.

Enabling robust cryptography in CPS remains a large challenge according to several presenters. The distribution, updating, and revocation of cryptographic keys presents a particular challenge as many CPS utilize hardware with certain constraints—such as amount of power, bandwidth, and processing capabilities—that cyber-only systems do not have.

Likewise, usability needs to be more widely considered and improved in CPS. Without good usability, security measures may be bypassed or users may become inattentive to systems that may need immediate attention. While good usability is an important trait in cyber-only systems, it becomes critical to systems where poor usability can lead to inattention or accidental misuse in a system with physical impacts.

Virtual models are essential to the design and construction/assembly of reliable CPS, which are often so complex that testing of prototypes is either prohibitively expensive or impractical. Key characteristics of the models that are needed are robustness, with accurate representation of the full suite of properties of a CPS and the complicated environments in which they must operate, potential for use in verification and validation, and interoperability, allowing the combined use of multiple models or component modules. One of the unique impacts of not having usable, robust virtual models of CPS is that many systems rarely get patched as there is no acceptable way to foretell what the results of a patch may be to the system, and the possibility of the system being unavailable for some time has financial and safety/health risks that outweigh the risk of not patching.

Failures of data integrity in CPS can result in tangible, real-world consequences not commonly seen in cyber-only systems. For example, in the military, loss of CPS data integrity can result in inaccurate or erroneous resource deployment, weapons targeting, etc. Likewise, a loss of integrity of data in networked medical devices can cause significant harm to a patient. The possible physical impacts present a significant, often unique challenge when designing cybersecurity for CPS.

Finally, some of the most difficult challenges are those that are not only technical in nature, but include business and policy aspects. For instance, producing a persuasive business case for increased cybersecurity efforts can be difficult. Sometimes the likelihood of an incident can seem small so small that the cost of the appropriate countermeasures seem too expensive. This points to the need for research and development of both better measures for verifiable assurance in components and systems and also cyber-economic tools to help assess the costs associated with the range of potential cybersecurity incidents.

In addition to the unique cybersecurity requirements for most CPS for cryptography and modeling, there is also a requirement to keep systems available despite ongoing security incidents. For example, turning parts of the Smart Grid off in order to thwart an attack is not possible. In some CPS, there is a need to be able to detect small amounts of malicious code within a very large overall amount of code. In the semi-conductor example presented, there was a need to find 1,000 – 2,000 lines of malicious code in 1 to 2 million lines of code, and disposing of "suspect" semi-conductors would have had an enormous financial impact for the company.

There is also a need with CPS to consider potential impacts of cybersecurity incidents slightly differently. In every CPS, there is a physical action or reaction that is controlled by a cyber system. In many cases, there are many possible physical actions (e.g., modern airplanes have many physical actions/reactions controlled by cyber systems). This means that virtually every CPS has a health, safety, and environmental (HSE) impact. This makes potential impacts of cybersecurity incidents relatively high compared to traditional cyber-only systems. And since many of these systems have a high availability need, the choices for possible mitigations can be limited. These HSE impacts can also create a need for additional requirements often not needed for cyber-only systems, such as the requirement to coordinate with local emergency responders in case of incidents. HSE impacts need to be considered very carefully when identifying cybersecurity requirements for CPS.

The Computer Security Division plans to use the results of this workshop to inform future research, publications, and outreach activities in the area of CPS cybersecurity.

The agenda for the workshop, complete with links to abstracts and slide presentations, may be found at http://csrc.nist.gov/news_events/cps-workshop/cps-workshop-agenda_04-03-2012.pdf.

www.ingramcontent.com/pod-product-compliance
Lightning Source LLC
Chambersburg PA
CBHW081435170526
45166CB00008B/2210